Student Workbook for *Organic* CHEMISTRY

JUDIT BEAGLE • LUCAS BEAGLE

Kendall Hunt
publishing company

Cover ©Shutterstock.com

Kendall Hunt
publishing company

www.kendallhunt.com
Send all inquiries to:
4050 Westmark Drive
Dubuque, IA 52004-1840

ISBN: 979-8-7657-0967-2

Published in the United States of America

Contents

Preface

Why we wrote this workbook.

We decided to write this manual after years of seeing where students struggle in organic chemistry. Many students think organic is difficult because there are so many reactions to memorize, this is not true. There are a lot of reactions, you do have to memorize them to be successful. As with any course there is a certain amount of discipline specific knowledge or lingo that you simply must know to understand the material. The chemical reactions are simply the vocabulary that is communicated in organic chemistry. However, we find that this is not what holds back students from being successful, it typically is the fundamentals.

We find that some things from general chemistry are not fresh in our memories, sometimes it was because it was a long time since we saw them (maybe even several semesters) or they were never really developed in our minds as fully as we thought in general chemistry (acids and bases). However, it is not just information from general chemistry, it is some of the fundamentals that are revisited throughout the entire course that are missed (nomenclature, etc.).

So how should you use this workbook? It should be a compendium and not a replacement for your text or notes. It is targeted practice on those areas that we have identified as trouble hotspots for students. Each section is written independently from another and covers the whole topic with information and practice. It can be helpful for both organic I and organic II students as it covers material learned in both courses that tend to be forgotten along the way.

We hope you find this helpful and welcome feedback or other areas that need to be addressed.

Judit Beagle
Lucas Beagle

Application of General Chemistry Concepts

1

What you will learn:

- Formal charge
- Hybridization
- Resonance structures
- Acid and base definitions
- Acid and base strength in terms of pK_a
- Structure effect on acid strength
- Acid–base equilibrium

1.1 INTRODUCTION

In the following sections we have compiled the areas of your general chemistry knowledge that seem to be the most troublesome for students in the organic chemistry classroom. Don't worry if you look at these topics and feel a little lost, it's okay it is understood that you had these in a previous semester and may not remember them completely. You may not have even had a great handle on some of these topics when you saw them last time. When it comes to acids and bases it is often that the concept is not understood until after you take organic chemistry and see how these important molecules work from a mechanistic aspect. With practice these concepts will begin to come back and you will remaster them quickly.

1.2 IMPORTANT DEFINITIONS

- **Formal charge:** Indicated the number of electrons formally assigned to an atom in a molecule
- **Resonance structures/resonance contributors:** A set of Lewis structures that collectively describe the electronic bonding of a single polyatomic species
- **Resonance hybrid:** The average of two or more resonance structures
- **Brønsted–Lowry acid:** Any substance that can act as a proton donor.
- **Brønsted–Lowry base:** Any substance that can act as a proton acceptor.
- **Conjugate acid:** Formed from a base by adding a hydrogen
- **Conjugate base:** Formed from an acid by removing the most acidic hydrogen
- **Lewis acid:** Any substance that can act as an electron acceptor.
- **Lewis base:** Any substance that can act as an electron donor.
- **Electrophile:** electron acceptor (Lewis acid), centers that are electron poor.
- **Nucleophile:** electron donor (Lewis base), centers that are electron rich.

1.3 FORMAL CHARGE

Formal charge and oxidation number were presented to you in your general chemistry course, these two concepts both accomplish the same task, the counting of electrons. Oxidation number treated electron counting as if we were dealing with ionic bonds between atoms, and formal charge treats them much more as if they were covalent. While both topics are useful, we will focus on the idea of formal charge as a method to formally account for the electrons on a specific atom in a molecule.

To begin we need to know how many valence electrons are on the neutral atoms. If we use the traditional names for columns on the periodic table (Groups 1A-8A) we can easily tell how many valence electrons should be present. For example, N (nitrogen) is in group 5A (a member of the pnictogen family), therefore it will have five valence electrons. This idea will work very well for us, as in organic chemistry we are only interested in the main group elements. Once we know the electrons on a neutral atom, we can then determine the formal charge on that atom. A positive formal charge fundamentally means we have less electrons formally on the atom than protons while a negative formal charge would indicate we have more electrons on the atoms than protons to balance out the charge. To count how many electrons are formally on an atom, we take the number of lone pair electrons and add half the number of bonded electrons the atoms have, then we subtract that combined value from the expected number of electrons of a neutral atom (Figure 1-1).

$$FC = [\#\ of\ valence\ electrons] - [\left(\frac{1}{2}\ of\ bonded\ electrons\right) + (lone\ pair\ electrons)$$

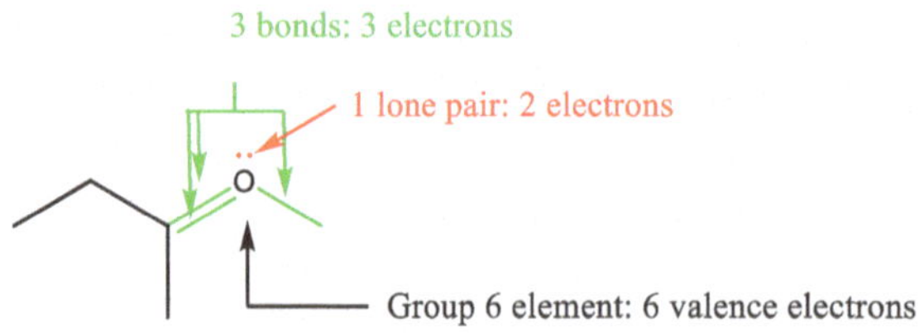

6 valence electrons
– 2 electrons (from lone pair)
– 3 electrons (half of bonded electrons)

+1 formal charge on oxygen

Figure 1-1: ***Formal charge formula and example of an oxonium ion.***

Practice

1.3.1. For the following molecules draw in all missing lone pairs and assign any formal charges.

a)

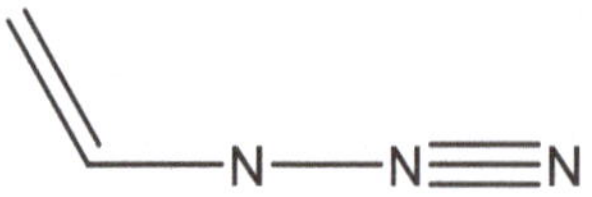

b)

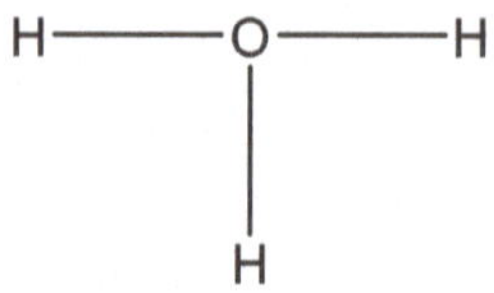

c)

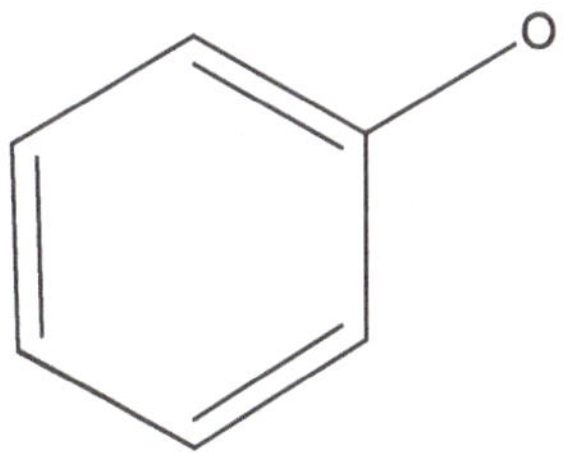

d)

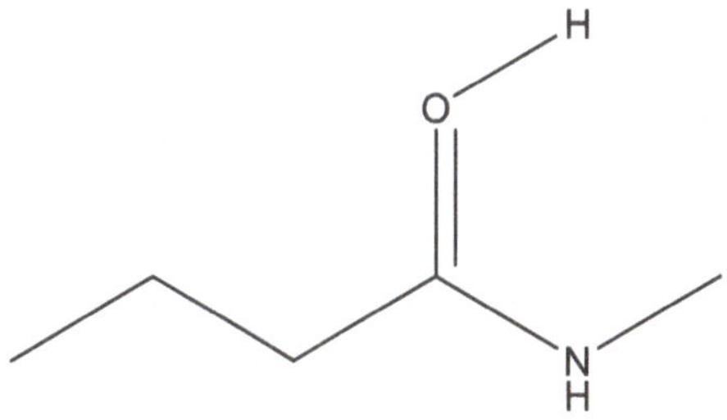

e)

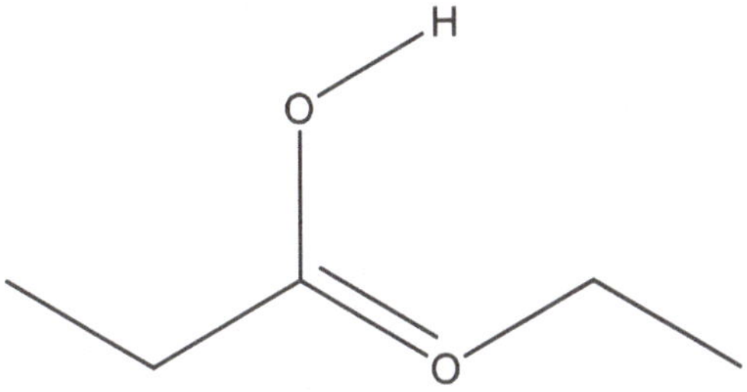

f)

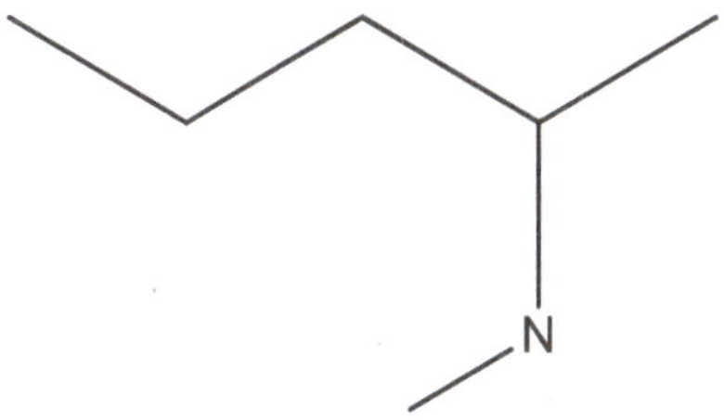

1.4 HYBRIDIZATION

Hybridization is one of those topics that is best practiced. It is one of the pillar concepts of general chemistry that impacts the fundamentals of organic chemistry (and ties in intimately with your knowledge of VSEPR theory). For in-depth details and explanations of why hybridization occurs you should reference either a general chemistry or organic chemistry textbook. We are interested in a practical understanding of hybridization and how to identify hybridization. In a simple definition, hybridization is a mixing of the s and p orbitals to allow degenerate (equivalent energy level) bonding orbitals to rationalize the bonding patterns we see in chemistry (think methane and the four degenerate bonding orbitals that carbon exhibits). In organic chemistry we will focus only on the hybridization of the s and p orbitals, we have three levels of hybridization we are interested in: sp^3, sp^2, and sp. What do we mean when we say sp^3, sp^2, and sp: this indicates the extent of the mixing of the orbitals (Figure 1-2).

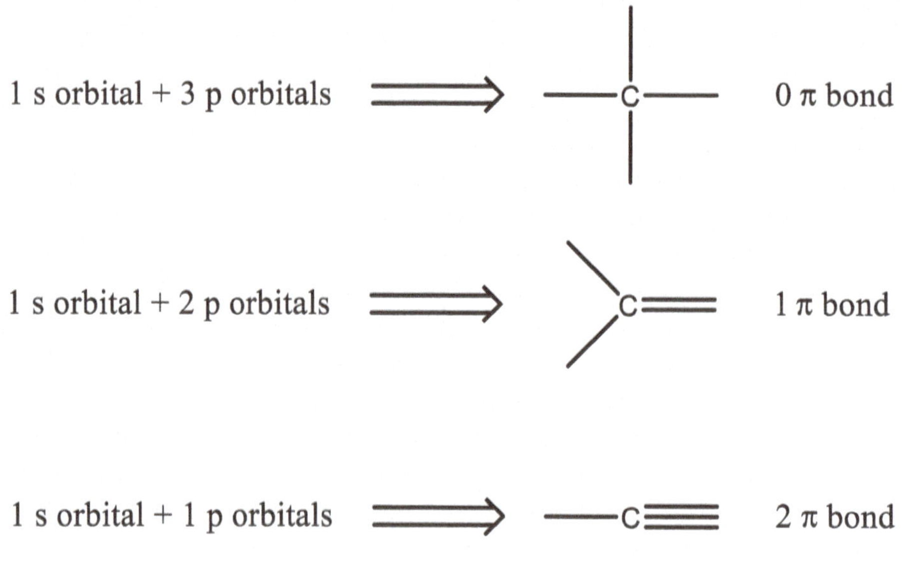

Figure 1-2: Example of what hybridization can occur.

As was stated earlier, in organic chemistry we are only interested in the s and p orbitals, so we are only looking at the four possible orbitals that we need to take into account when looking at molecules. The assumption is that all orbitals are hybridized (see your textbook for a detailed explanation of why this occurs) in order to have the lowest energy possible for the molecule. If this is the case we can easily assign the hybridization of any atom in a molecule. If we start with the assumption that all orbitals are hybridized, then we start by assuming any atom is sp^3 (one s orbital mixed with 3 p orbitals) hybridized. The next assumption is that any pi-bond must occur in a p orbital, and it will not mix to form a hybridized orbital. What does this mean, it means that if we have a pi-bond, we subtract 1 p orbital from our mixing. This means that we have one p orbital that we cannot count, we now only have two p orbitals to mix with the s orbital resulting in a sp^2 hybridized atom (Figure 1-3). If we have two pi-bonds to an atom we then need to use two p orbitals to make the pi-bonds, leaving us only 1 p orbital to mix with the s orbital resulting in a sp hybridized atom.

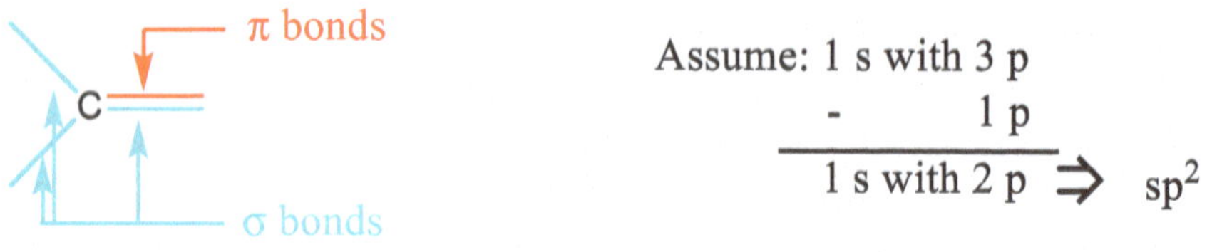

Figure 1-3: Example of what hybridization can occur.

Some common things that cause students problems involve other atoms than carbon, lone pairs, and hydrogen. Firstly, we can assign hybridization to any row two heteroatoms (and many more but we focus on these) in a molecule, not just carbon using the previous guidelines, in the exercises you will commonly be asked to assign hybridization for carbon, nitrogen, and oxygen. Lone pairs are also in hybridized orbitals! Students commonly will ignore the lone pairs and assign oxygen as sp, or nitrogen as sp^2 because they fail to account for the lone pairs. Hydrogen is not hybridized, it only has the 1s orbital and therefore cannot form hybridized orbitals.

Practice

1.4.1. Determine the hybridization of the indicated atoms in the following molecules.

a)

b)

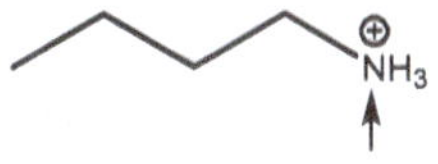

c)

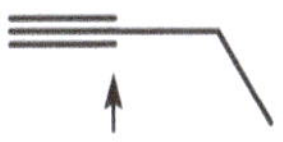

d)

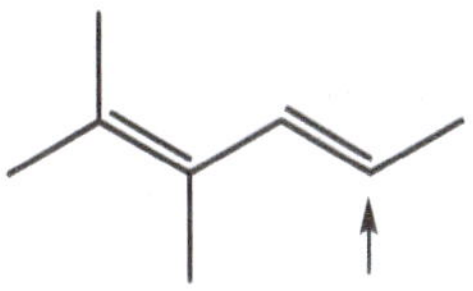

e)

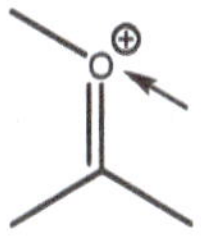

f)

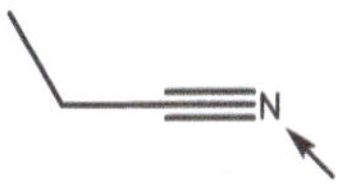

g)

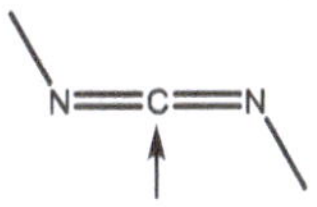

h)

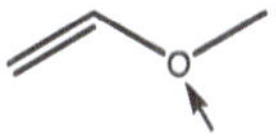

i)

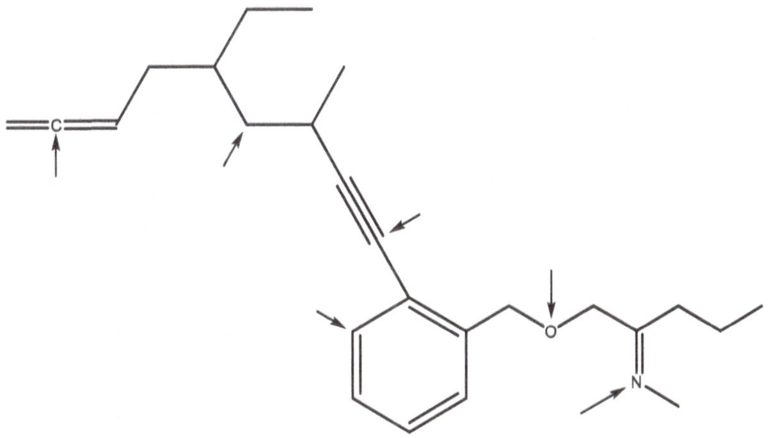

1.5 RESONANCE

Resonance may be one of the hardest concepts for students to master during general or organic chemistry, the whole idea is about equivalent imaginary structures. This makes it incredibly difficult, however if we follow some simple rules and look at a few basic examples we will be prepared to tackle any resonance problem we may come across. We will see that the concept of resonance will help us explain why chemicals react the way they do as we try to account for the observed reactivity. This becomes especially important in the second part of organic chemistry where we see electrophilic aromatic substitution. The observed reactivity can only be explained by looking at the individual resonance contributors of these substituted benzene rings.

What is resonance? We can think of it as how a molecule spreads around electrical disturbances. Why is this important, with what we know in chemistry is that if we have a lone pair of electrons it is a high energy situation. Anytime that we can distribute a high energy occurrence around a molecule it would spread out the energy, so each atom had less of an impact from that occurrence. Think about this in simpler terms, if you had a couch (this is your electrical disturbance) and you needed to move it. Would sharing the load with a couple of people make it easier to bear the burden of the couch, of course it would. In the chemical world lone pairs, single or unpaired electrons (radicals) and positive charges (empty p orbitals) are all more stable if the burden of that disturbance is shared around the molecule were possible. In the next section we will see what makes for valid resonance structures and what are invalid resonance structures.

Let's start with some basics, if we were to draw benzene (Figure 1-4) we could draw two possible arrangements of the internal, alternating double bonds. You might ask yourself, which is the right one, is one better than the other? Well they are the same (kind of), yet the bonds are in different places, how can this be okay? They are each individual resonance structures (or resonance contributors) providing

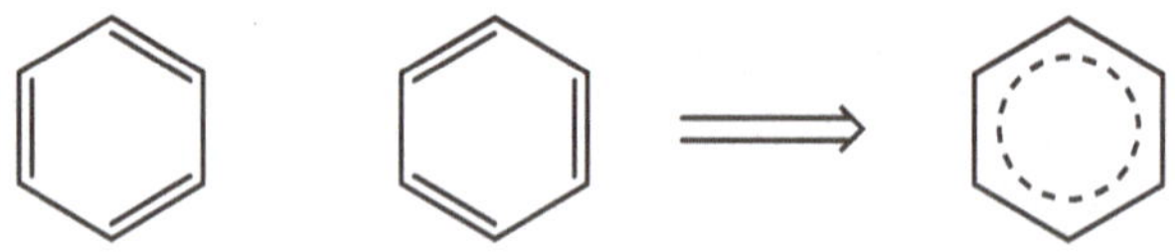

1 2 Resonance hybrid

Figure 1-4: ***Individual resonance contributors and the resonance hybrid of benzene.***

the overall state of the molecule. Each contributor is kind of like a photo or snapshot in time of the molecule as the electrons are moving around the molecule. We use these contributors to help understand all the possible electron configurations or arrangements when delocalization or resonance occurs. We can now understand that there are many individual resonance structures that can be visualized when we have electrons that can delocalize, but what does the molecule look like overall? That overall picture is the resonance hybrid; it is the comprehensive picture of all the individual resonance structures. These resonance hybrids often have a bizarre appearance as they don't show the electrical disturbance on any particular atom but spread across sections of the atom, this is indicated by replacing the resonating pi-bonds with dashed lines.

While the rules that govern resonance are fairly straightforward, when coupled with actually moving the electrons around the molecule it can feel confusing. Below you will find the basic rules for resonance.

1.6 BASIC RULES FOR RESONANCE

Rule 1 – Individual resonance contributors are imaginary and not real! They only serve to help us understand how the electrons are moving.

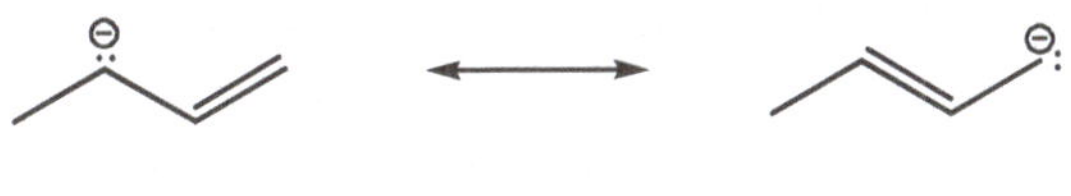

Rule 2 – Resonance structures only differ in their placement of pi-bonds andtheir electrical disturbance (lone pairs, radical (unpaired) electron, or positive charge).

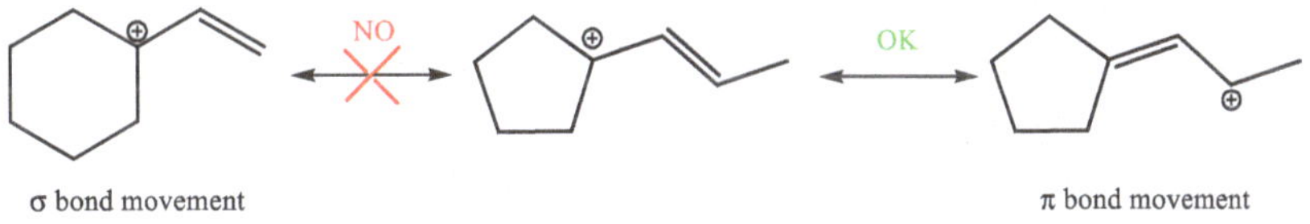

Rule 3 – Resonance forms do not have to be equivalent to each other.

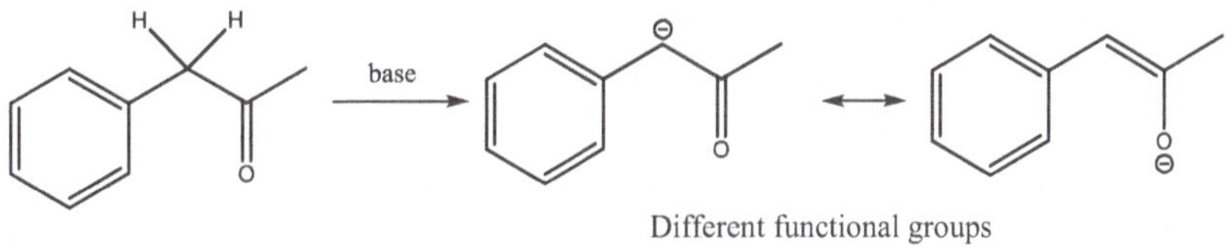

Rule 4 – Resonance forms must not violate valency (No 5 bonded carbons! No 2 bonded hydrogens!)

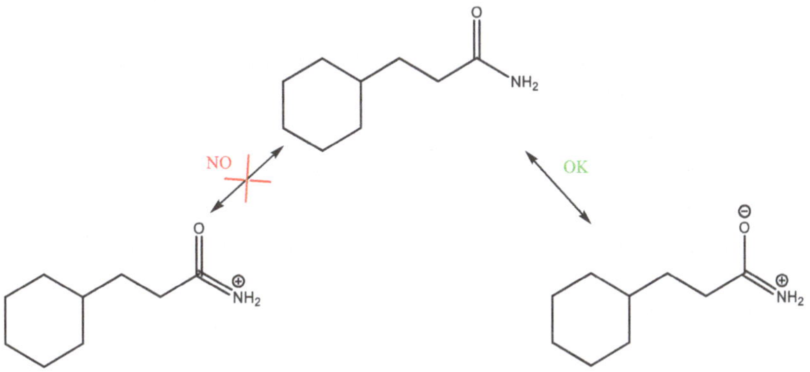

Now that we have these basic rules under our belts let's look at how we show the three basic electrical disturbances resonate or delocalize around the molecule. For resonance to occur, you must have your electrical disturbance and immediately alpha (next to) to the disturbance a pi-system (Figure 1-5). The amount of resonance structures you can draw will be limited to the extent of conjugation of the pi-system. This resonance can occur with any pi-system, we commonly encounter double bonds (i.e., C=C, C=O, C=N), one of the pi bonds involved in a triple bond or an aryl group. A common mistake is to show a disturbance resonating where it cannot (Figure 1-6). A common mistake occurs when the pi-system is further than the alpha position from the disturbance. Alternatively, the next common mistake is an exaggeration of the conjugation of the pi-system (i.e., there was an sp^3 carbon between the pi-systems that were ignored).

If you are feeling like resonance can be a difficult concept, it can be. Mastering resonance takes time, practice, questions and an awful lot of mistakes to get it right!

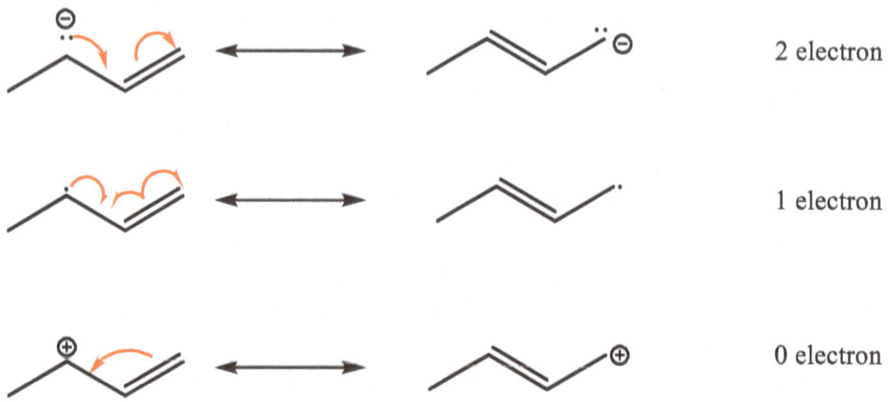

Figure 1-5: ***Resonance of lone pair (2 electrons), radicals (1 electrons), and positive charges (0 electrons). There are two important concepts to remember when moving electrons, the arrow starts at the electron rich area and ends at the electron poor; secondly, use single-headed arrows to move one electron and a double-headed arrow to move two electrons.***

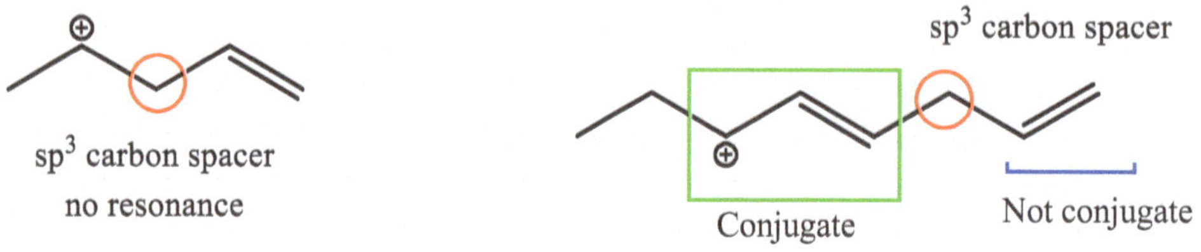

Figure 1-6: ***Some examples of common mistakes in resonance.***

Practice

1.6.1. Draw the resonance structures of the molecules below.

a)

b)

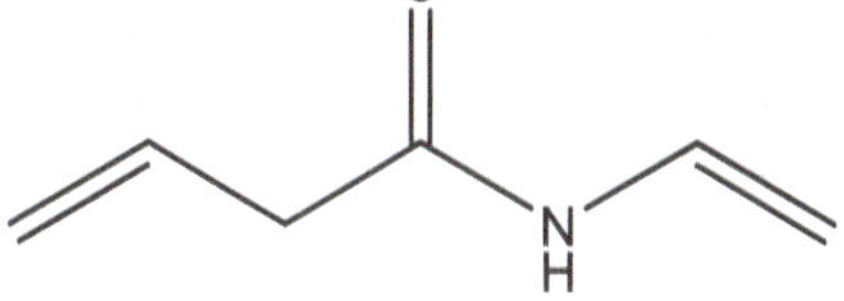

c)

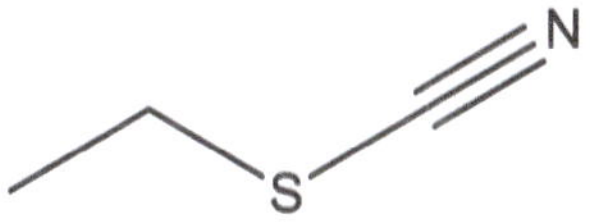

d)

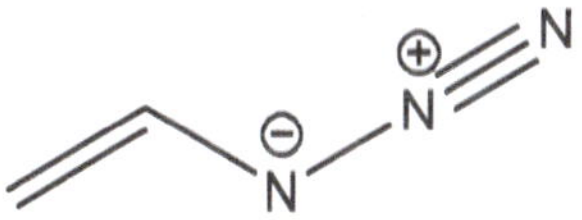

e)

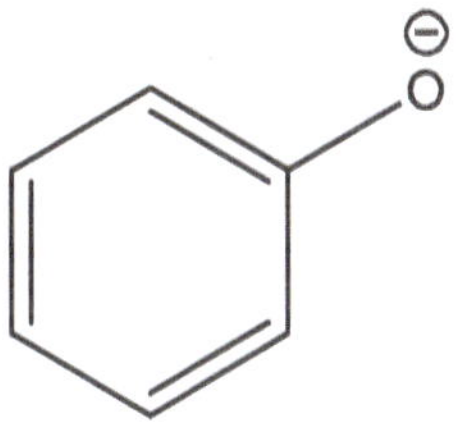

f)

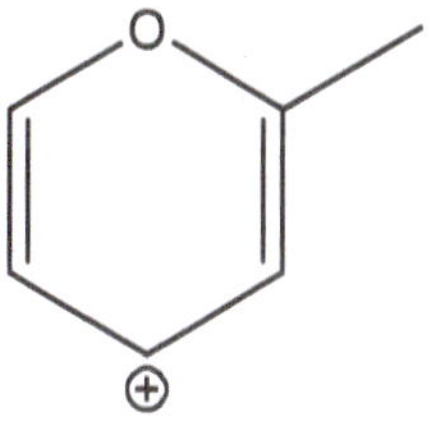

g)

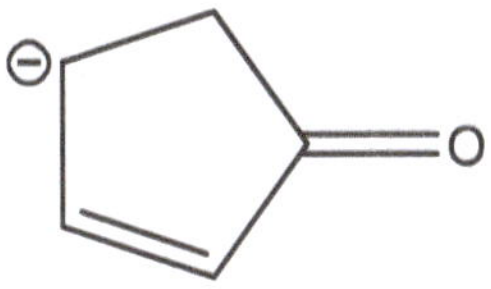

h)

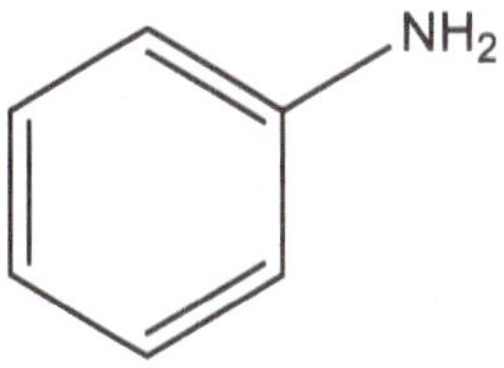

1.7 ACIDS AND BASES

Acid and base reactions are some of the most common reactions that occur, they are often involved as the steps of many of the mechanisms throughout organic chemistry. While the concepts can be easily stated in definitions, mastery is not always that simple. Much of the confusion with acids and bases is the nomenclature. All the words we use to talk about acid and base reactions deal with the acid, for example, protonation, these all lead to a misconception of what goes on in an acid and base chemistry. In general chemistry you focused on Brønsted–Lowry acid and bases, this is still the case in organic chemistry; however, now we will add Lewis acids and bases (Figure 1-7).

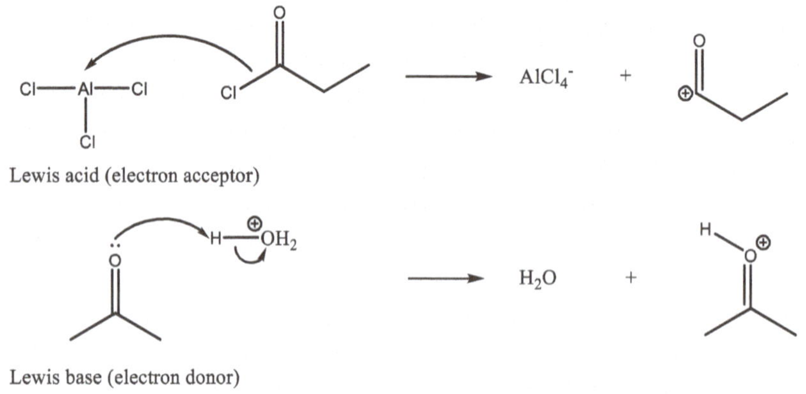

Figure 1-7: ***Example of Lewis acids in organic reactions.***

Until organic chemistry there has not been a need to talk about what is mechanistically happening in these reactions. When we view the acid and base reaction in terms of the mechanism, we gain a better understanding of what is actually occurring at the chemical level (Figure 1-8). Notice the directionality of the arrows, it is the base that is doing the action not the acid. The base removes a proton from the acid, for us this is a conceptual change, as we have always heard the acid protonates other chemicals. The acid still has vital importance in the reaction, and we need the acid to have a labile bond to the abstractable hydrogen. In fact, the strength of an acid is measured in how easily the bond to the abstracted hydrogen is broken. However, it is the base that is removing the hydrogen form the acid and forming a bond with the hydrogen in the process. When we view in these reactions in terms of how the electrons flow, acid and base chemistry becomes much easier to understand.

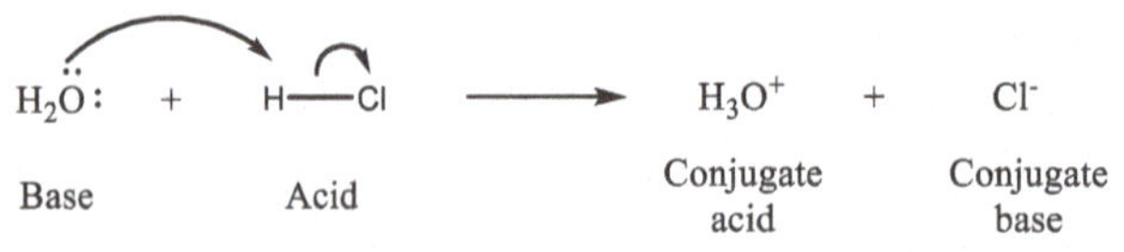

Figure 1-8: ***A simple acid and base reaction.***

One of the most troubling concepts for students is how do I determine the strength of the acid. When thinking about the strength of the acid, we actually want to look at the conjugate base. Again, we are not looking at the acid itself for this property, it seems really strange as we have been conditioned to think only of the acid. **The strength of an acid is directly related to the stability of the conjugate base.** This is one of the most important things to remember when trying to compare chemicals to rank their acidic strength (or many other properties in chemistry, nucleophilicity, basicity, etc.). We commonly are interested in three factors of the conjugate base when doing comparisons: (1) resonance, (2) ionizability, and (3) electronegativity and inductive effects. Why do we look at these factors, because when the base abstracts the proton, it leaves behind a lone pair of electrons. This lone pair of electrons wants to undergo basic attack, if these lone pair electrons are stabilized by lowering their energy, it will be unfavorable to react regenerating the initial acid. The more stabilizing factors the conjugate base possesses the stronger the acid (Figure 1-9). Sometimes multiple factors occur to increase the stability. In the next section we will discuss a concrete value to be used to quantify acid strength, if you examine the conjugate base you can get a reasonable qualitative answer.

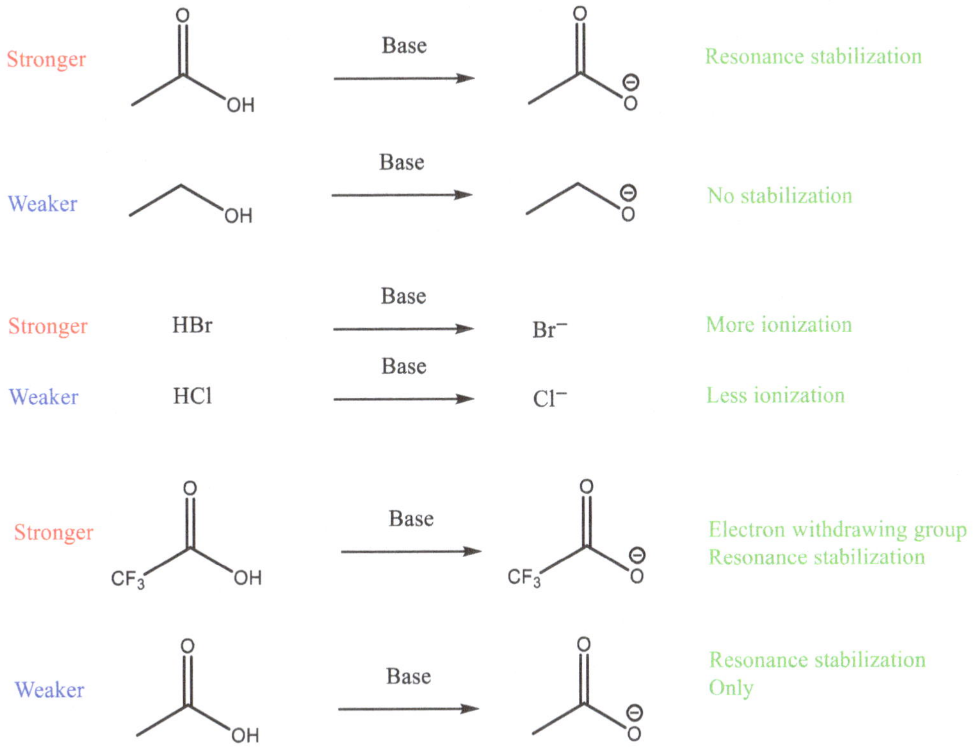

Figure 1-9: ***Examples of stabilizing forces of conjugate bases that correlate to stronger acid behavior.***

Practice

1.7.1. In each of the following pairs, circle the stronger acid.

a)

b)

H-Cl H-F

c)

NH_3 NH_4Cl

d)

H_2O $HO^{\ominus}$

1.7.2. Rank the following acid in order of increasing acidity.

a)

b)

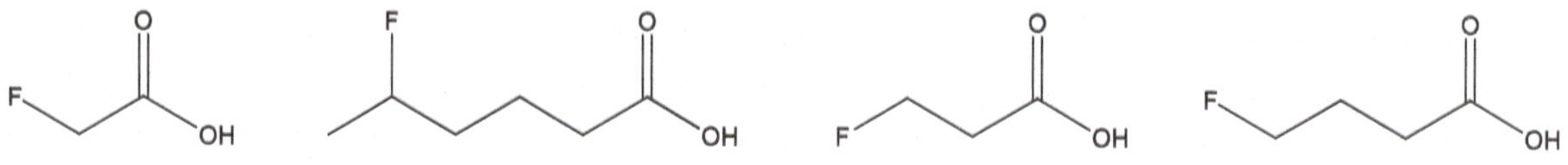

c)

d)

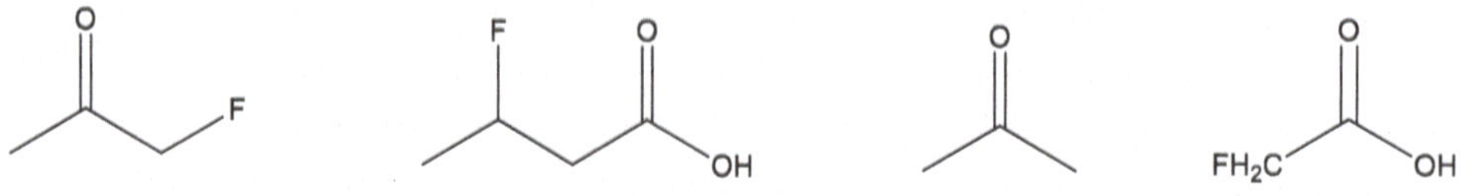

e)

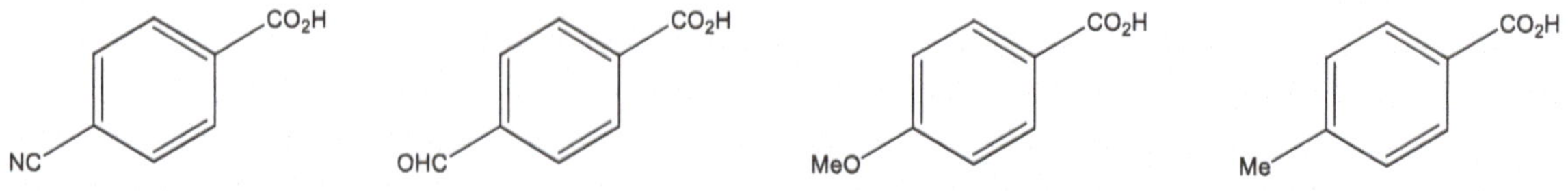

pK_a and Acid–Base Equilibrium

From general chemistry we remember that there were some fundamentals of predicting a chemical reaction, one of those was equilibrium (kinetics and thermodynamics were the others). In general chemistry you spent a great deal of time writing and solving equilibrium expressions, the reason why you did this was not just to get a value, but to get a judgment. This judgment you arrived at fundamentally was how well the chemical reaction works. Since acid/base reactions are in equilibrium, we can investigate how well they work by looking at the equilibrium expression. We can use the acid–base equilibrium to determine how well an acid (or conversely a base) acts like an acid (acid strength).

In acid–base reactions that equilibrium expression is given a unique name called acid dissociation expression and given a subscript a to the equilibrium constant, K_a (Figure 1-10). Conversely you can invert the expression and get the base association constant, K_b, as well to understand how well the base behaves as a base. What are we really looking at when we see an acid dissociation expression, we are really seeing how labile that acid–hydrogen bond is, that is the strength of the acid.

$$HA\ (aq) + H_2O\ (l) \circledR H_3O^+(aq) + A^-(aq)$$

$$K_a = \frac{[H_3O^+][A^-]}{[HA]}$$

Figure 1-10: Acid dissociation expression for an acid–base reaction.

As a refresher, K_a values greater than 1 means it is a strong base and values less than 1 are a weak base. However, these values are extremely large and make no rational sense. We saw this same idea in general chemistry when we looked at [H_3O+], these values must be treated with a negative log function to give understandable and usable numbers, the pH values. While pH told us about the concentration of hydronium in solution, it did not tell us anything about the strength of the acid molecule. We can transform the K_a values in the same manner to get pK_a values, these are the concrete numbers that allow us a general chemistry style comparison of two acids (Figure 1-11). The values for pK_a read contrary to K_a values, the stronger the acid the lower the number, the higher the number the weaker the acid.

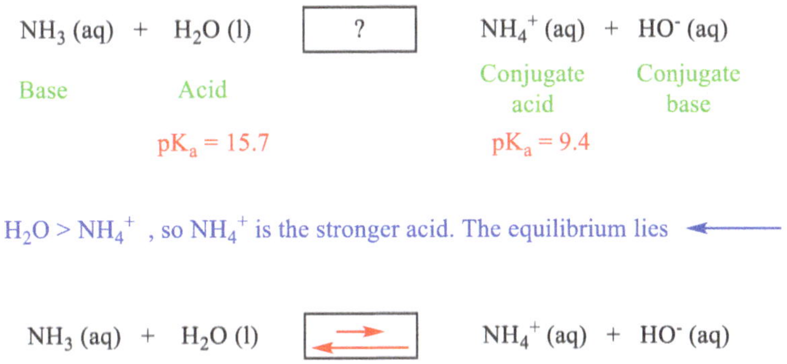

Figure 1-11: Determining the equilibrium position of an acid–base reaction. The side of the equation with the lowest pK_a value determines the equilibrium position. Values for pK_a can be found in your text book or other reference.

Practice

1.7.3. Using the pK_a values indicate the favored direction of the acid–base equilibrium.

a)

$CH_3CH_2O^-$ + HF [?] CH_3CH_2OH + NaF

b)

$H-C\equiv C-H$ + HO^- [?] H_2O + $H-C\equiv C^{\ominus}$

c)

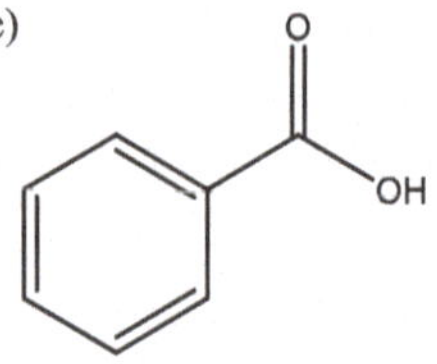
+ NH_3 [?] NH_4^+ +
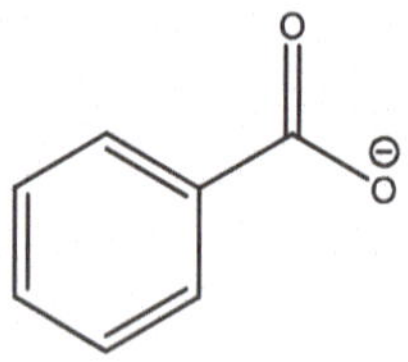

d)

NH_4^+ + H_2O [?] NH_3 + H_3O^+

e)

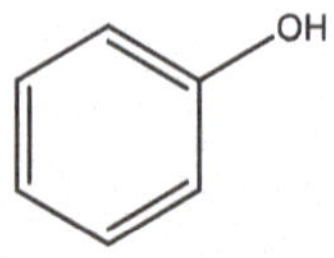
+ $NaNH_2$ [?] NH_3 +
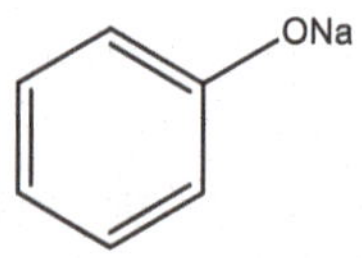

1.8 CONCLUSION

General chemistry made us feel comfortable because we were always dealing with numbers, you got a value, it was right or wrong. You didn't always need to think of the impact of that value, in organic chemistry we find the opposite is true. Being able to get to a number is not important, being able to use the number is important. This often makes organic chemistry more difficult on students who like to crank through a formula; you will find that you will need to think about the output more, you will need to develop what is called "organic chemistry intuition." Not always having concrete values to make determinations (such as acids and bases without memorizing large tables of values associated with each compound), but instead having to look at molecules and building a profile of that molecule and what is occurring. You will see later in the course, there are often many ways to solve a problem, there is no cookie cutter approach and many offshoots can occur.

Name: ________________________________ Date: ____________

INTEGRATED PROBLEMS

1.1. For the following molecules draw in all missing lone pairs and assign any formal charges.

a)

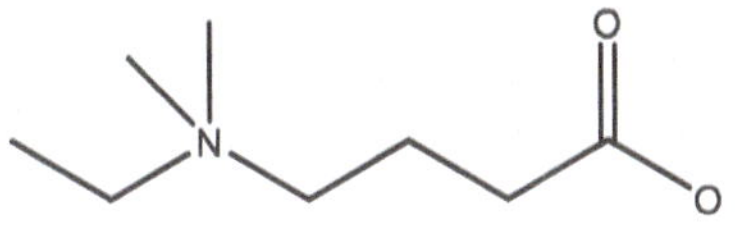

b)

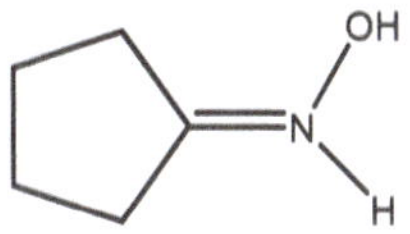

c)

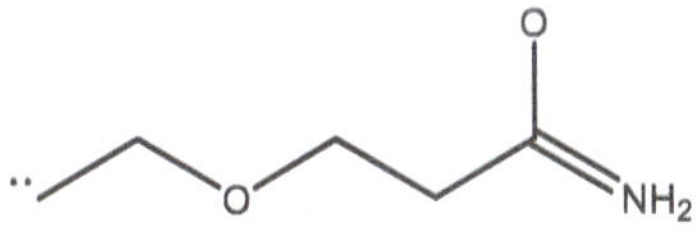

d)

e)

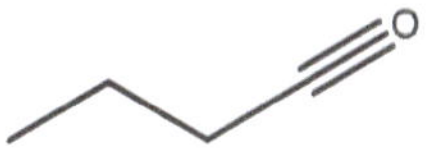

f)

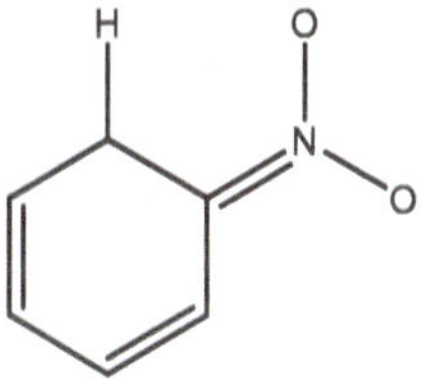

1.2. Determine the hybridization of the indicated atoms in the following molecules.

a)

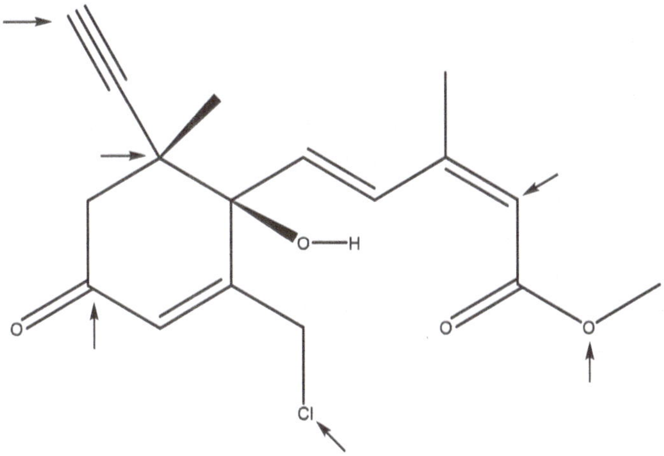

b)

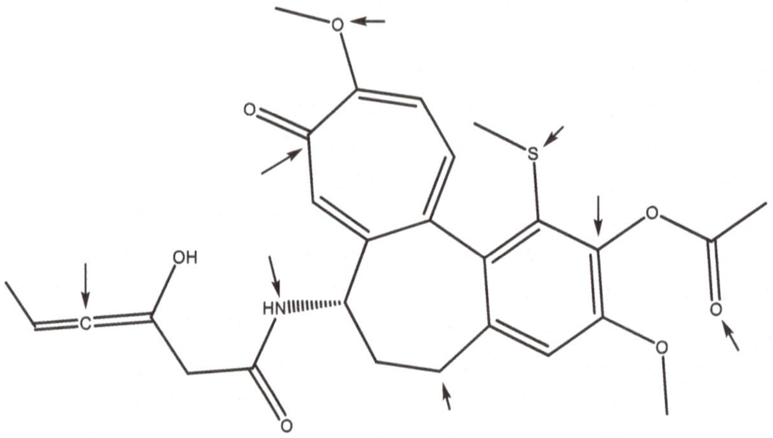

1.3. Draw the resonance structures of the molecules below.

a)

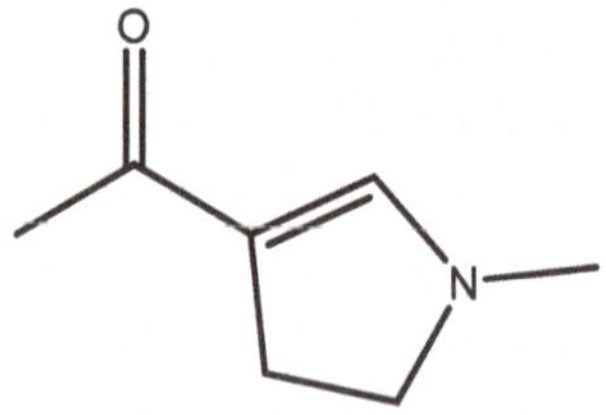

b)

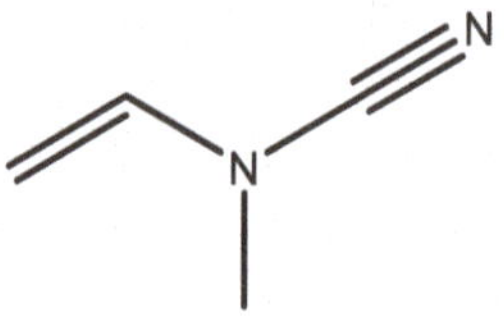

c)

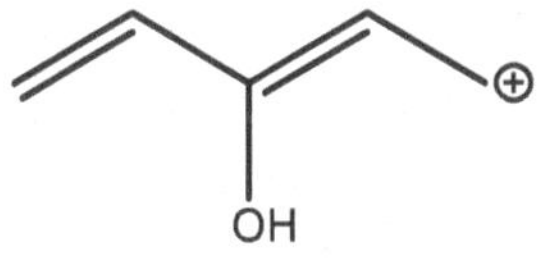

d)

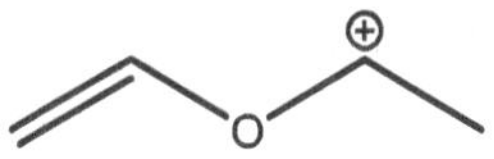

e)

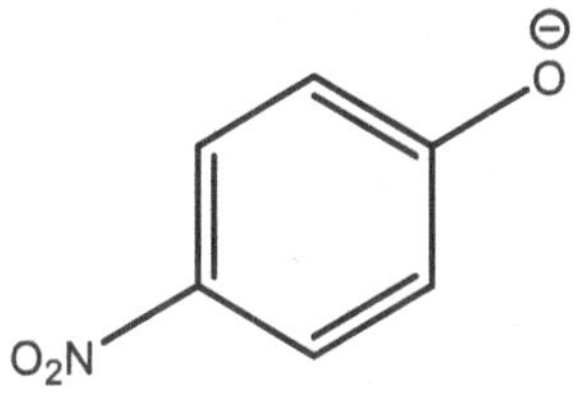

f)

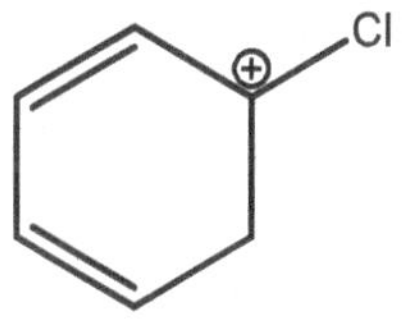

1.4. Using the pK_a values complete the following equations and indicate the favored direction of the acid–base equilibrium.

a) **LEFT RIGHT**

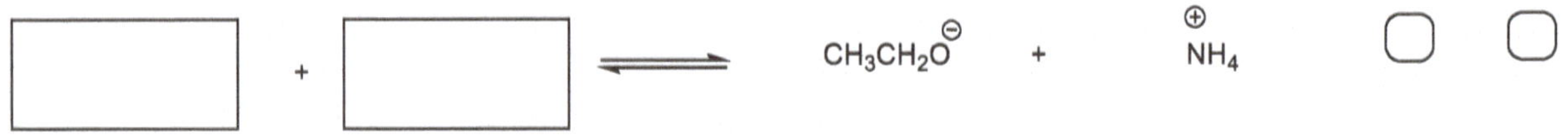

b)

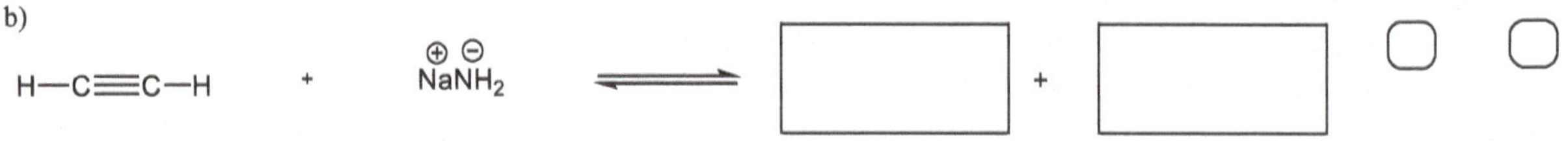

c)

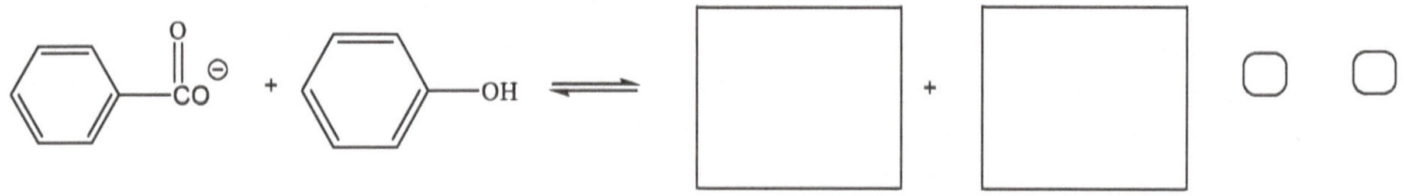

d)

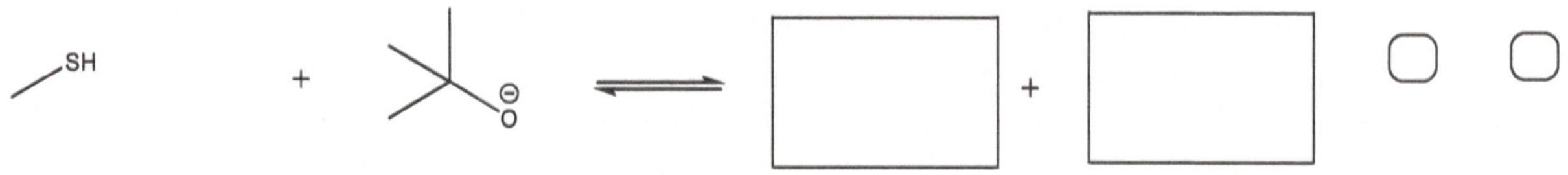

e)

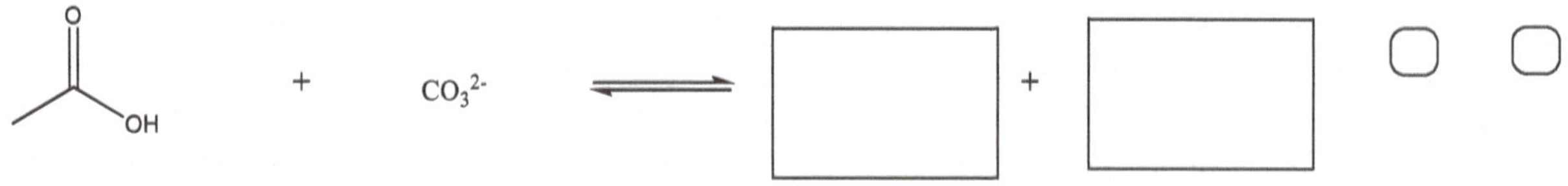

f)

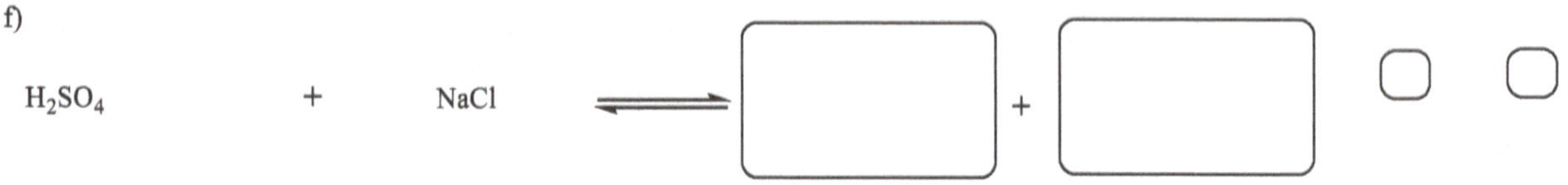

g)

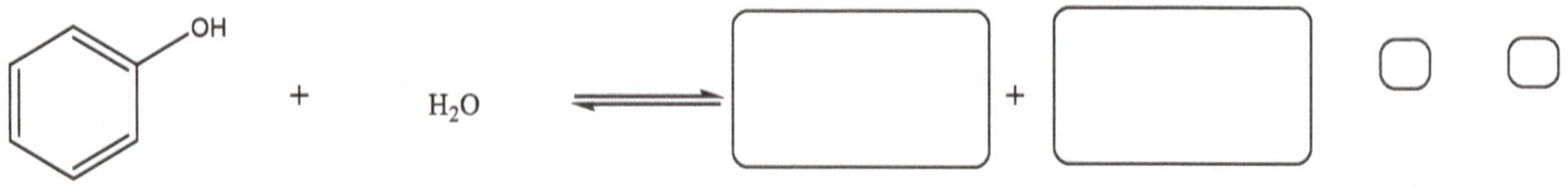

h)

Drawing Organic Molecules

2

What you will learn:

Draw and interpret organic molecules in different representations:

- Lewis structure
- Condensed formula
- Line-angle structure
- Newman projection
- Fisher projection
- Chair configuration

2.1 INTRODUCTION

Organic molecules can be drawn in different representations, depending on what information we would like to convey. Some representations are two dimensional, while others are three dimensional and give us better understanding how the molecule really looks like. These representations are useful, when we asses stability, reactivity, and intermolecular forces (which lead to the determination of physical properties such as boiling point or melting point).

2.2 IMPORTANT DEFINITIONS

- **Lewis structure:** A representation where all atoms, bonds, and lone pairs are shown.
- **Condensed formula:** A representation of the molecule as a line of text. All atoms are shown, but omits the vertical bonds and most or all horizontal single bonds.
- **Line-angle structure:** A representation, where none of the carbons and hydrogens attached to them are shown. All other atoms are shown, and hydrogens attached to other atoms are also shown.
- **Newman projection:** A two-dimensional representation of a three-dimensional molecule used to analyze the conformation of a specific bond.
- **Fisher projection:** A two-dimensional representation of a three-dimensional molecule used to represent primarily carbohydrates.
- **Chair conformation:** Three-dimensional representation of a cyclohexane ring.

2.3 LEWIS STRUCTURE

While Lewis structures are rarely used in organic chemistry, because organic molecules in general contain a lot of atoms, it is important that you are confident in generating Lewis structures. It will help you to better understand the representations to follow. You have learned in general chemistry the basic rules of drawing Lewis structures. Let us review it now.

Lewis structures show each atom and its position in the molecule using its chemical symbol. Lines are drawn between atoms that are bonded to one another. Excess electrons that form lone pairs are represented as pairs of dots, and are placed next to the atoms.

There are many methods known for the construction of Lewis structures. While the basic electron counting method was useful in general chemistry (Figure 2-1), it can be overwhelming in organic chemistry, because of the huge number of total electrons.

C_2H_6O

total number of electrons = sum of valence electrons of each atom = 4 + 4 + 1 + 1 + 1 + 1 + 1 + 1 + 6 = 20

Step 1: Connect all atoms by a single bond

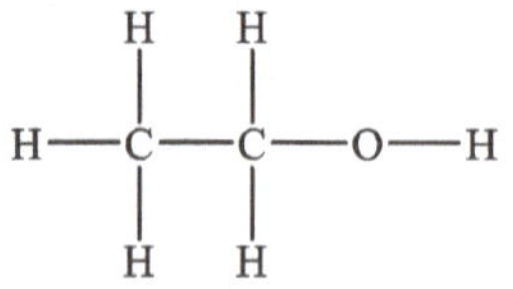

Step 2: If t is the total number of electrons and n the number of single bonds, t–2n electrons remain to be placed. t = 20, n = 8, therefore 20 – 16 = 4 electron remains. These should be placed as lone pairs: one pair of dots for each pair of electrons available. Lone pairs should initially be placed on outer atoms (other than hydrogen) until each outer atom has eight electrons in bonding pairs and lone pairs; extra lone pairs may then be placed on the central atom. When in doubt, lone pairs should be placed on more electronegative atoms first

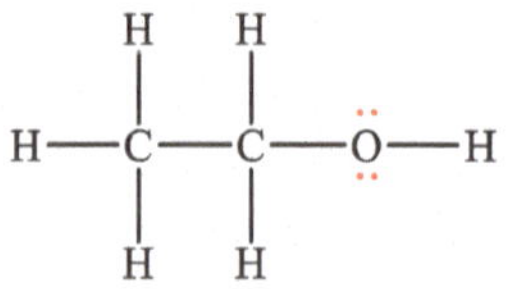

Each atom now has its octet, so the structure is correct.

Figure 2-1: ***Drawing Lewis structure using the electron counting method learned in general chemistry.***

A simplified method can be used in organic chemistry. As shown in Figure 2-2.

C_2H_6O

Step 1: Connect all carbon atoms in a single line with single bonds. If you have any other atom, connect them with single bonds as well.

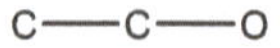

Step 2: Add the hydrogens to the atoms to achieve octet. Start with carbon atoms.

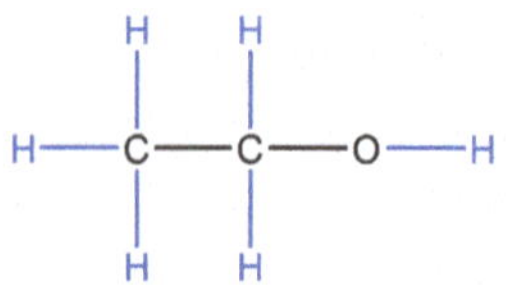

Step 3: If any atom (carbon or otherwise) does not have octet after placing all hydrogens, add lone pairs or double bonds.

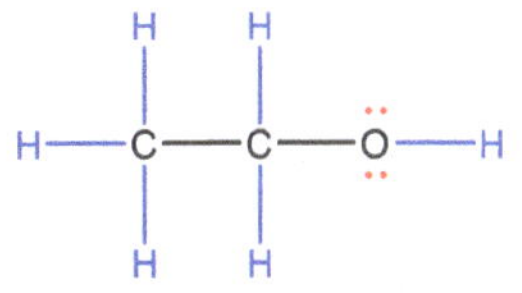

Figure 2-2: ***Using the simplified method to draw Lewis structures.***

Please note that even for a small structure as the one above, this is not the only correct Lewis structure that can be drawn. These different structures are called constitutional isomers, where the molecular formula is the same, but the order in which the atoms are connected (connectivity) is different.

Practice

2.3.1. Draw possible Lewis structures that does not include ring or charged atoms. Make sure you draw all your lone pairs.

a. C_2H_3N

b. C_3H_6O

c. C_4H_9Cl

d. C_5H_{12}

2.4 CONDENSED FORMULA

As you have seen in the previous section, Lewis structures can get quite large. A simplified representation is the condensed formula, where most of the bonds are not drawn out, but all atoms are shown. This representation lacks any three-dimensional information about the molecule. To create a condensed formula from a Lewis structure, you read the Lewis structure from left to right and write down the carbon first, then everything that is connected to it. Keep doing this until you reach the end of the chain. If you have a branch, you draw that bond to the appropriate carbon and continue until the end. Two examples are shown (Figure 2-3).

Converting Lewis structure to condensed formula – without branch

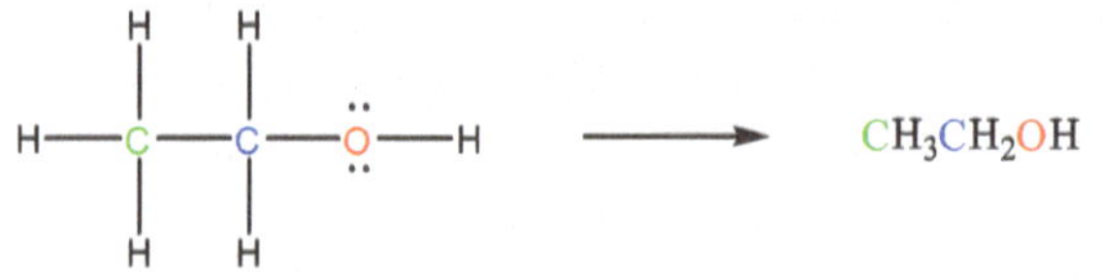

Remember: The blue carbon and the red oxygen are the ones connected, not the hydrogen and oxygen

Converting Lewis structure to condensed formula – with branch

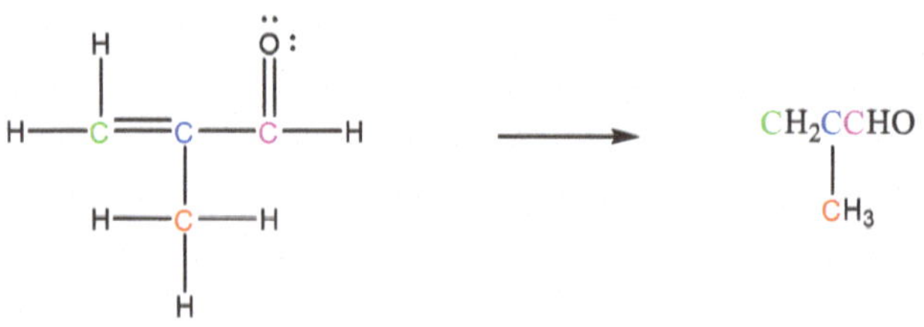

Remember: The purple carbon has both the oxygens and hydrogens connected, not the oxygen to the hydrogen

Figure 2-3: ***Converting Lewis structures to condensed formulas.***

Practice

2.4.1. Draw possible condensed formulas that does not include ring or charged atoms.

a. C_2H_3N

b. C_3H_6O

c. C_4H_9Cl

d. C_5H_{12}

2.5 LINE-ANGLE STRUCTURE

While the condensed structure simplifies the drawing of organic molecules it is still rather tedious. As noted above, the disadvantage to the condensed structural formula is that it ignores skeletal connectivity and three-dimensional structural information (both concepts are important as discussed in later chapters pertaining to functional groups and stereochemistry); in addition, the condensed structure often leads to mistakes as drawing the structure of each atom must be accounted for in the molecule. Line-angle structure, also commonly called skeletal structure, is a method highlighting the connectivity of the carbon skeleton. The line-angle structure is the most common way to convey chemical information in organic chemistry.

The real power of the line angle lies in the simplicity of the form. There are three simple rules when writing a line-angle structure: (1) carbons are represented as vertices or endpoints; (2) hydrogens are implicit to the structure and are not drawn unless attached to a heteroatom; (3) heteroatoms are always represented as their chemical symbol. Below are more detailed examples as well as an example (Figure 2-4). Students often find it most useful to relate these to their condensed structure counterparts when learning how to draw a line-angle structure. Three-dimensionality of the structure is achieved by the addition of wedges and dashes to the carbon skeleton to maintain the VSEPR geometry learned in the general chemistry course. The main carbon skeleton is drawn in the plane of the paper. A wedge denotes that a group is being elevated above the plane of the paper and a dash indicates a group is being demoted below the plane of the paper.

Carbons are represented as vertices or end points of line segments, the vertices/endpoints are connected with a line showing a bond between carbon atoms. If there is more than one bond shared between atoms a second line segment is added between the two atoms.

$CH_3CH_2CH_2CH_2CH_3$

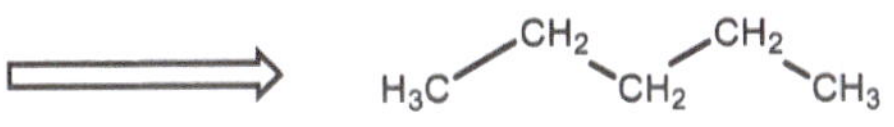

Hydrogens are not drawn on the skeleton of the structure, this allows for a cleaner looking structure and less opportunity to make a mistake while drawing the structure.

Heteroatoms (atoms that are not carbon or hydrogen) are always represented with a bond to the skeleton structure. Hydrogens are ALWAYS draw on heteroatoms if they occur.

$CH_3CH_2CH_2CH_2NH_2$

Figure 2-4: The rules for drawing molecules in line-angle formula.

Practice

2.5.1. Draw possible line angle structures that does not include ring or charged atoms.

a. C_2H_3N

b. C_3H_6O

c. C_4H_9Cl

d. C_5H_{12}

2.6 NEWMAN PROJECTION

Newman projections are an important way of visualizing the three-dimensional relationships of groups of atoms on a bond of interest. When viewing a Newman projection, it is important to realize you are looking down a bond between two atoms, most often two carbon atoms. The most difficult part of drawing a Newman projection is deciphering what a Newman project represents. We are trying to represent two ends of a bond with a face-on view of the bond. The "front" or "top" of the bond is represented by a center with three bonds projecting at 120° from each other. The "back" or "bottom" of the bond is represented in much the same way; however, there is a large circle drawn over the atom of interest representing the bond between the atoms; therefore, we do not see that "back" or "bottom" atom in the bond. We then take the two pieces of the model and set the "top" or "front" half over the "bottom" or "back" portion. See the example (Figure 2-5).

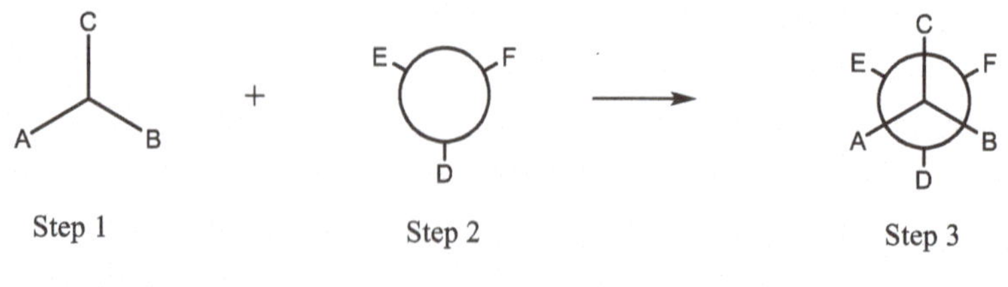

*Figure 2-5: **Stepwise method for drawing a Newman projection.***

One can think about identifying a bond between two atoms in a line-angle structure, then flipping the molecule 90° out from the paper to "view down" that bond. An important key to remember when translating the molecule through space is to keep groups oriented together, see example of the translation of a molecule from line-angle structure to Newman projection (Figure 2-6).

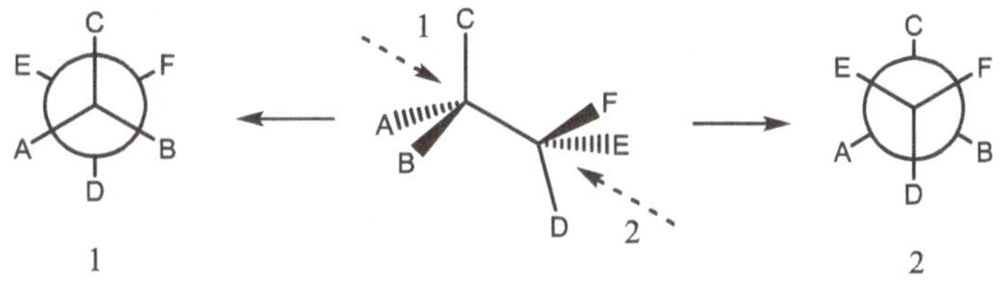

Figure 2-6: ***Conversion of Newman projections with line angle formula. The bond can be viewed from both directions.***

Another important use of the Newman projection is to represent the spatial considerations of the molecule. If visualizing two singly bonded atoms, there is free rotation about the bond, and therefore, there are many possible spatial relations of the groups (Figure 2-7).

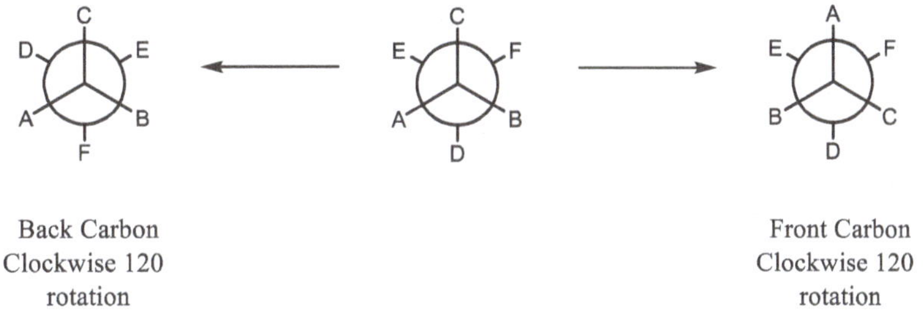

Figure 2-7: ***Rotation about the carbon–carbon bond showing the groups changing spatial orientation. Either the front or back carbon may be rotated.***

The above example showed the movement of the carbons by 120°, but the carbons do not have to move only in 120° increments. In organic chemistry we are concerned with a couple of important spatial relationships that occur during rotation of a Newman projection; these relationships can be seen in Figure 2-8. "Eclipsed" is used to denote the bonds aligned overtop each other and "staggered" is used to describe bonds that are rotated 60° from each other. We can also describe the relationship of two atoms or groups. By staggered, we can say that D and C are anti-periplanar because they are rotated 180° from each other, while we would say E and C are syn-periplanar because they are rotated 0° from each other.

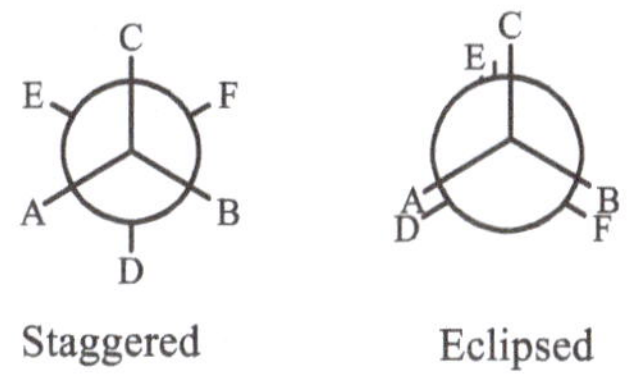

Figure 2-8: ***Important conformations of the Newman projections: staggered, and eclipsed conformers have different energy values associated due to the spatial relationship of the groups.***

Practice

2.6.1. Draw one of the Newman projections for the following molecules.

a. C2–C3 bond of $CH_3CH_2CH_2CH_2CH_3$.

b. C5–C4 bond of 4-methyl heptane.

2.6.2. Draw all the staggered and eclipsed conformers for the following molecules.

a. C3–C4 bond of 3,4-dimethyl hexane

b C2–C3 bond of 2-chloro-2-methylpentane

2.7 FISHER PROJECTIONS

Fisher projections are a useful way to exam larger molecules to study spatial relationships. Later we will learn how to assign R/S stereocenters to determine absolute stereochemistry that allows us to see similarities and differences among molecules of the same connectivity. Fisher projections are most commonly used when visualizing carbohydrates as they make it easy to identify differences quickly among the same carbohydrate class. For now, we are simply interested in understanding what makes a Fisher projection a unique way of displaying molecules.

Firstly, much like line-angle structure we are concerned with the carbon skeleton (or carbon backbone), and then we will put the groups that come off the backbone on later. Much like are line-angle structures; we will not be writing in many of our atoms, we will rely on vertices or more specifically in this case where lines cross to indicate the presence of a carbon atom. In the Newman projection we imagined identifying a bond of interest and translating it through space to view the bond, in the fisher projection we are going to take our line angle structure and straighten out the back bone to view it more easily. The vertical lines will be representing the carbon backbone, while horizontal lines are representing groups coming off the back bone. When orienting the molecule, the most oxidized (think back to your general chemistry days) end of the backbone goes to the top; therefore, the least oxidized end will go to the bottom (Figure 2-9). There are some spatial considerations that must go into the Fisher projection that are unique to help us understand where the groups are in space. Vertical lines are going behind the plane of the paper; whereas the horizontal lines are always considered to be above the plane of the paper; often this is denoted in the drawings as wedged or dashed lines, but not always!

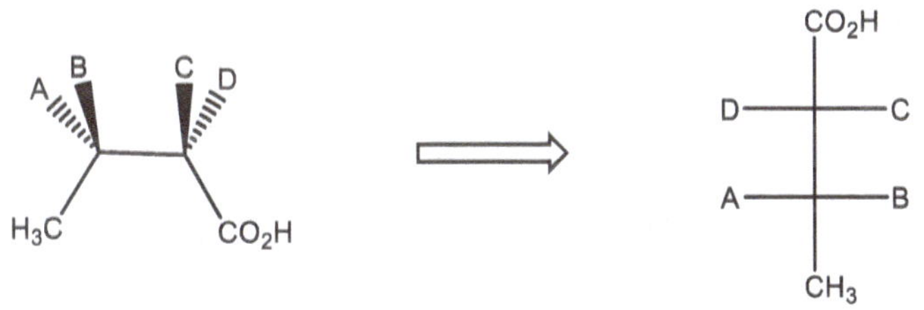

Figure 2-9: ***Conversion of molecules from line-angle structures to Fisher projections.***

Practice

2.7.1. Draw Fisher projections for the following molecules.

a. 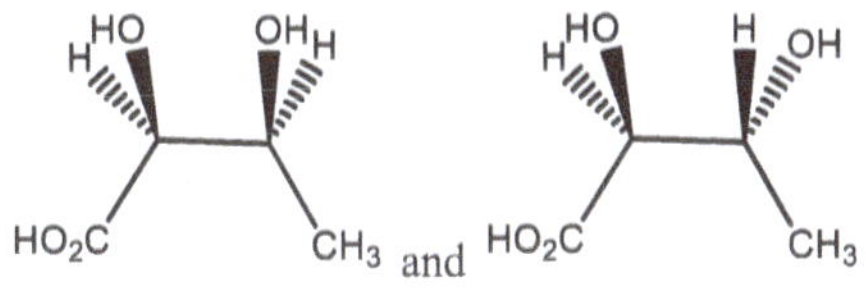

b. Draw the possible Fisher projections for all four carbon aldoses (erythrose and threose).

2.8 CHAIR CONFORMATION

Chair conformations are used to help understand the three-dimensional implications of a ring. This is much different from the acyclic drawings illustrated in the above text. The term chair conformation is used because its conformation adopted by a cyclohexane molecule looks like a chair. Why a chair-like conformation? The answer lies in the VSEPR theory: sp^3 hybridized centers have a tetrahedral shape and a 109.5° bond angle. For the 120° internal angles of a hexagon to become 109.5°, the molecule must rearrange through puckering and rotations to achieve the desired geometry. The hardest part of drawing a cyclohexane in the chair conformation is drawing the chair conformation, the below example will illustrate the correct way to draw the two forms of a cyclohexane ring (Figure 2-10).

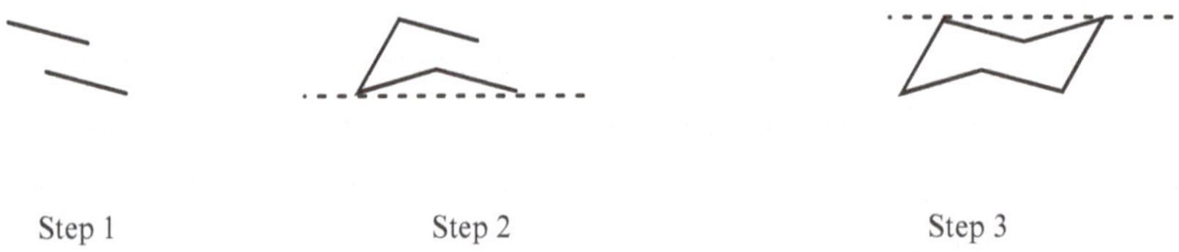

Figure 2-10: ***Stepwise drawing of the cyclohexane ring. Step 1—two parallel lines are drawn slightly offset. Step 2—the parallel lines are connected at a vertex parallel to the offset lines. Step 3—the last vertex is made between the offset lines parallel in height to the other offset line.***

Now we have the ring completed, however that only accounts for two of the four bonds that each carbon will have. The other two bonds are either away (equatorial) or above or below the plane (axial) of the ring. Each carbon has one axial and one equatorial bond to achieve the VSEPR geometry. Below you will see an example of how to draw the axial and equatorial positions are drawn, note that the axial is always following the point of the vertices that denotes the carbon (Figure 2-11).

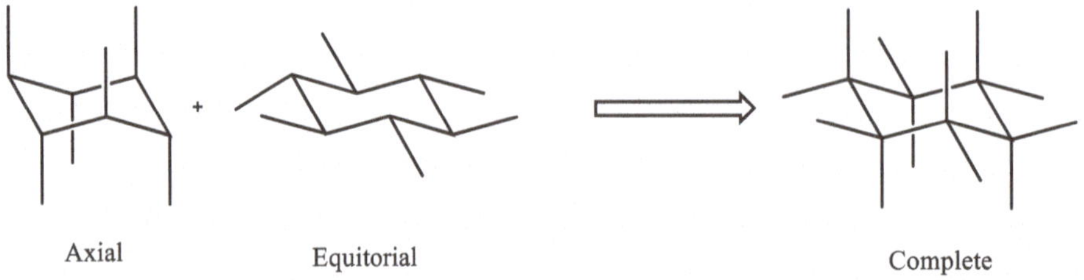

Figure 2-11: ***Axial (6) and equatorial (6) positions on a cyclohexane ring, each carbon contains one axial and one equatorial position.***

The last information to know about drawing rings is the ring flip. Ring flips occur because molecules are in constant motion and are rotating in solution. There are consequences for having groups on the ring that will often favor one form of the ring over another. As you know the larger the group, the more favored for it to be in the equatorial position. Below you will see an example of a ring flip of chlorocyclohexane, paying particular attention that if a group is axial in one conformer, it will be equatorial in the other, and if a group is equatorial in one conformer, it will be axial in the other (Figure 2-12).

Figure 2-12: ***Ring flip of chlorocyclohexane, the remaining axial and equatorial hydrogens are omitted for clarity.***

Practice

2.8.1. Draw one of the chair conformations for the following molecules.

a. 1-Chloro-2-methylcyclohexane

b. Isopropylcyclohexane.

c. 2-Ethyl-5-methyl-1-propylcyclohexane.

2.8.2. Draw ring flips for the following molecules

a. *t*-Butylcyclohexane

b. 1-Cholo-2-fluoro-cyclohexane

c. 1-Bromo-3-methyl-cyclohexane.

Name: ______________________________ Date: ____________

INTEGRATED PROBLEMS

2.1. Convert the following Lewis structures to condensed formulas!

a)

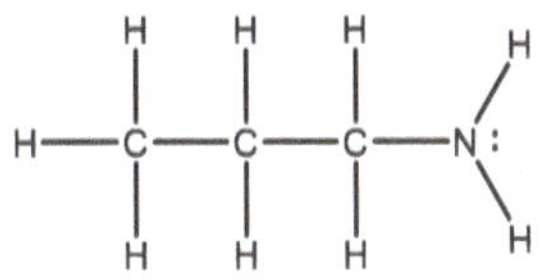

c)

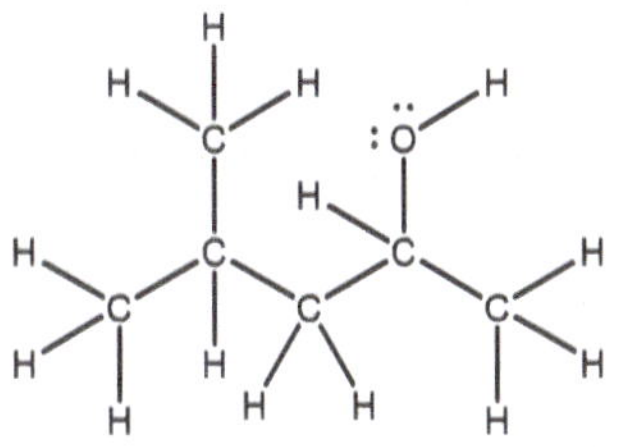

b)

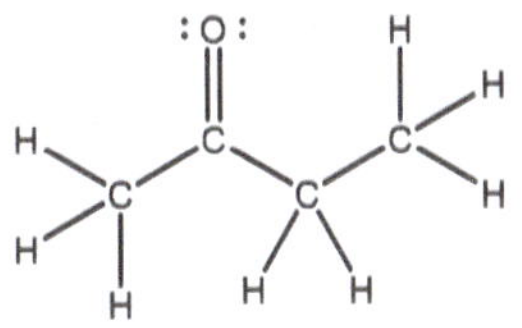

d)

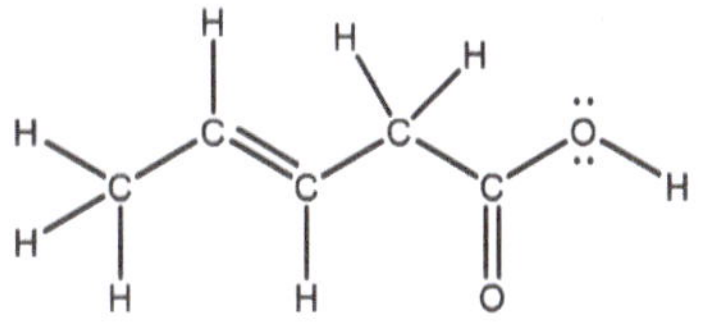

2.2. Convert the following condensed formulas to Lewis structures!

a. $CH_3CH_2CHClBr$

b. $CH_3CH_2CH_2CHCH_2CH_2CH_3$

c. $(CH_3)_2CHCH_2OH$

d. 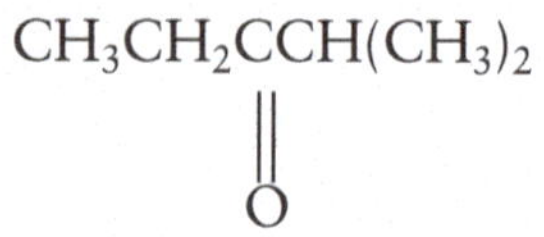

e. 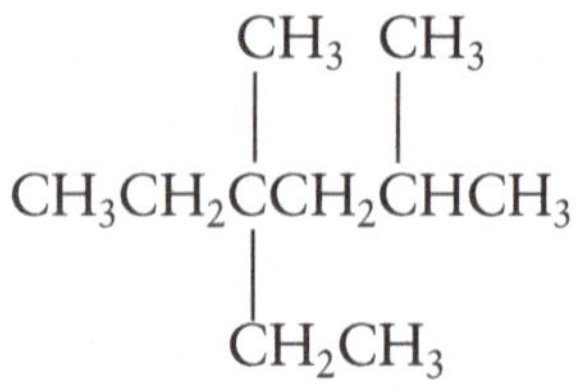

f. 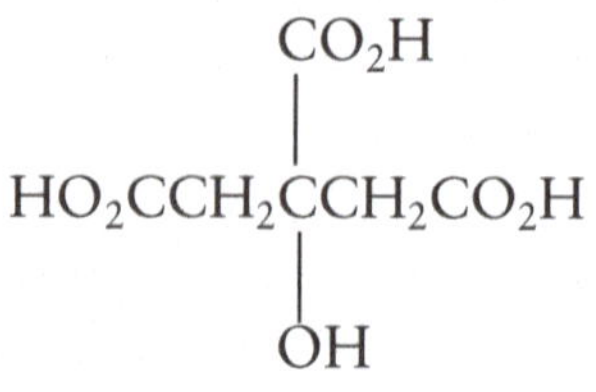

2.3. Draw possible Lewis structures, condensed formulas, and line angle structures which do not contain ring or charged atom for the following molecular formulas!

a. C_4H_9OH

b. C_6H_{12}

c. $C_5H_{10}Cl_2$

d. C_7H_{12}

e. $C_5H_{13}N$

2.4. Give the molecular formulas of the following molecules represented with their line angle structures!

a)

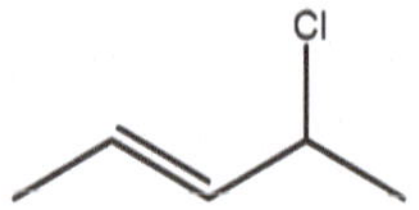

b)

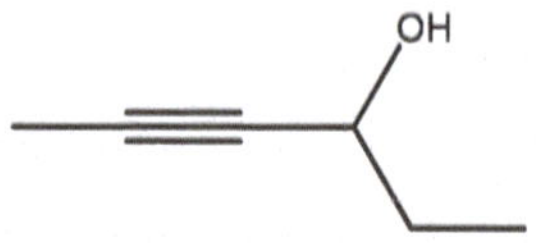

c)

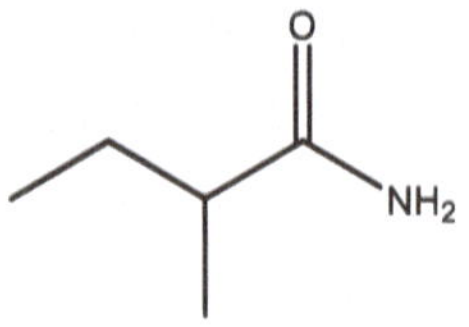

d)

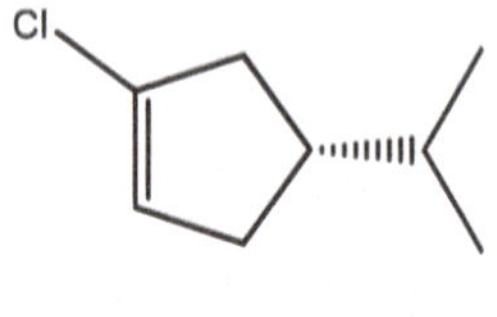

e)

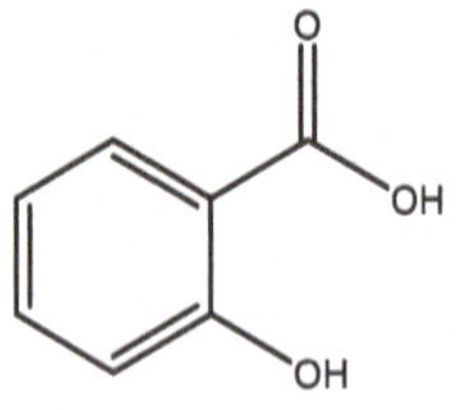

f)

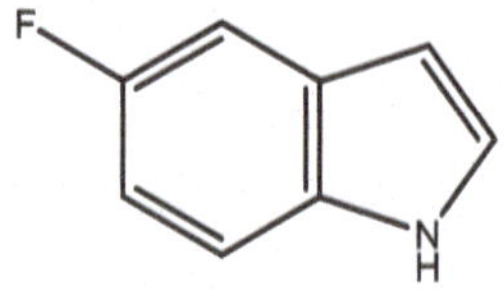

2.5. Draw the Newman projections of the following molecules looking down on the indicated bond(s)!

a)

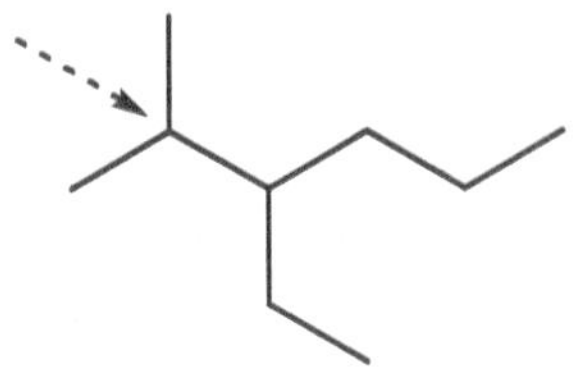

b)

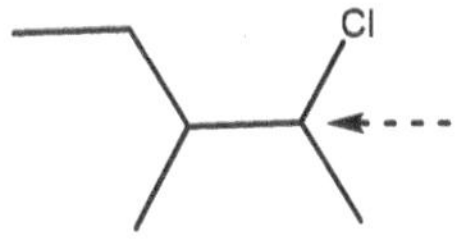

c)

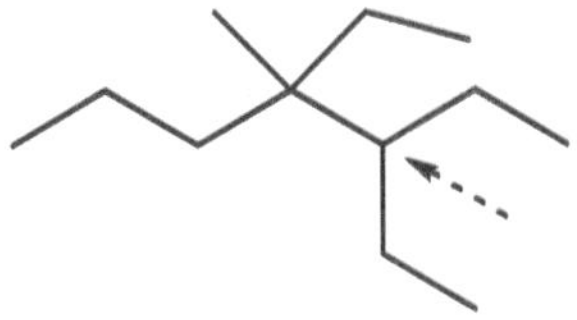

d)

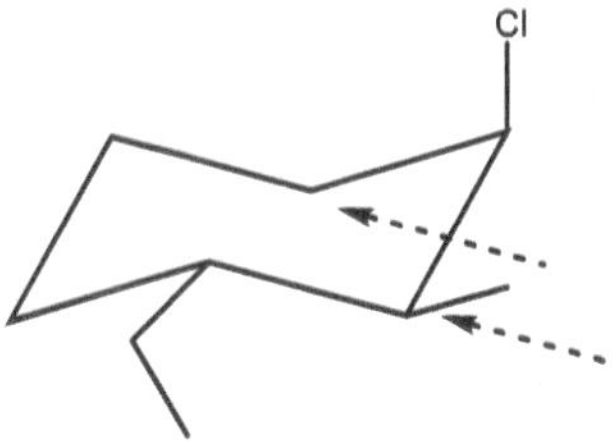

2.6. Convert the following Newman projections into line angle structures!

a)

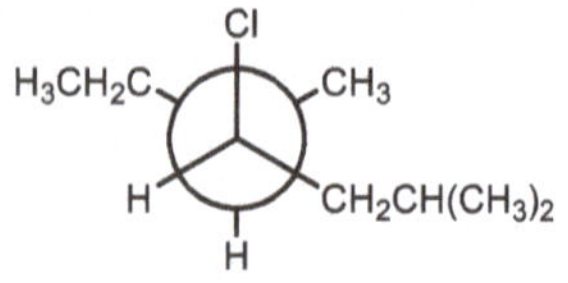

b)

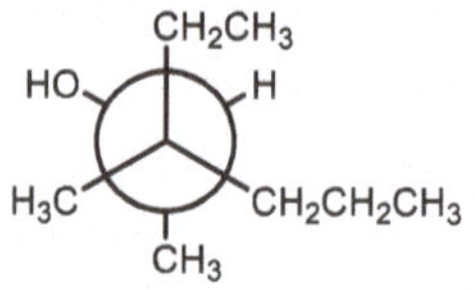

c)

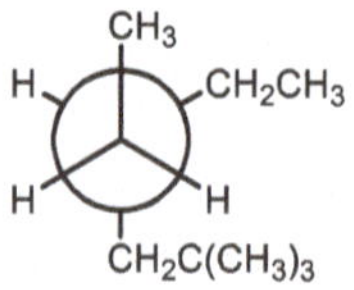

d)

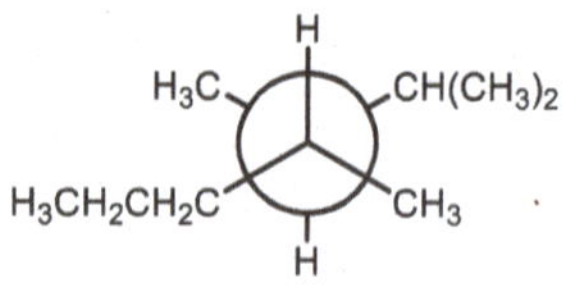

2.7. Draw the Fisher projection of the following molecules!

a)

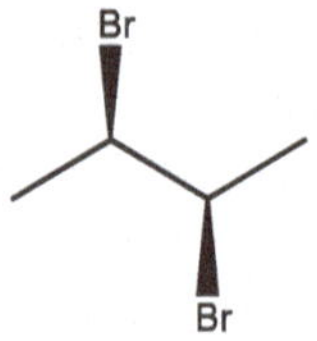

b)

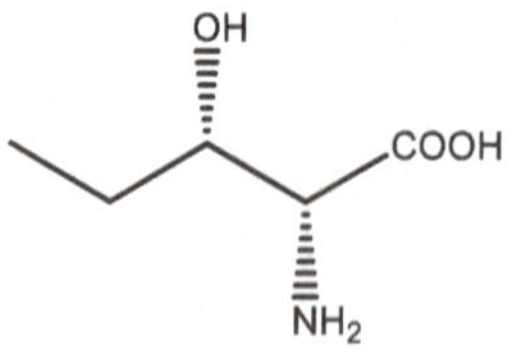

c)

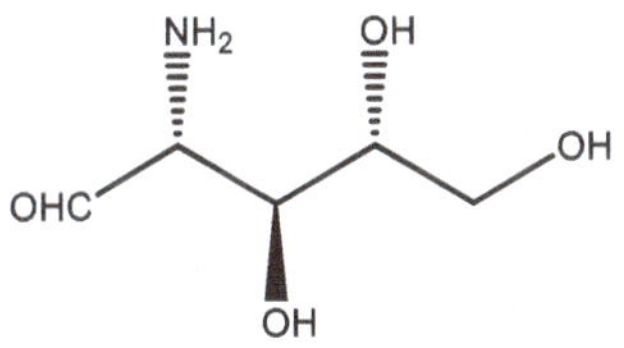

d)

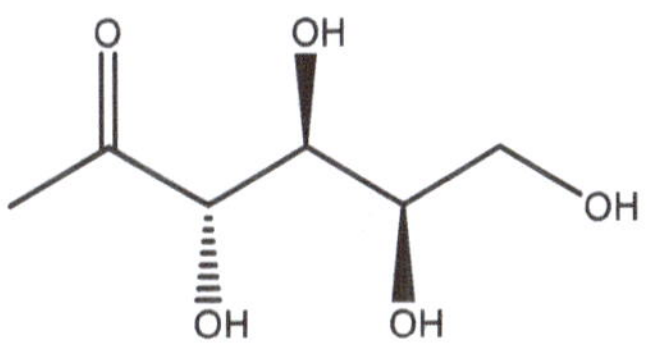

2.8. Draw both of the chair conformations for the following cyclohexane derivatives!

a)

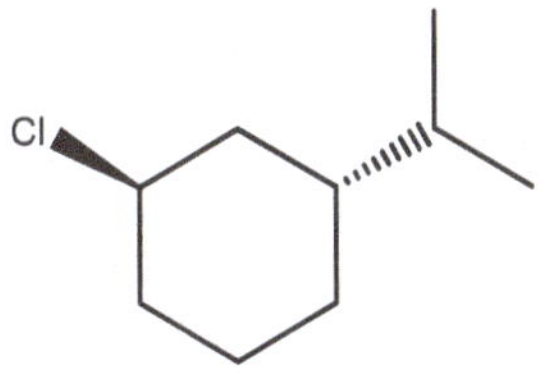

b)

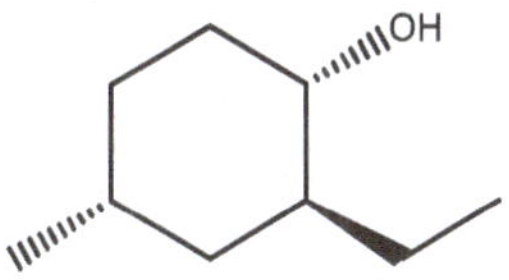

c)

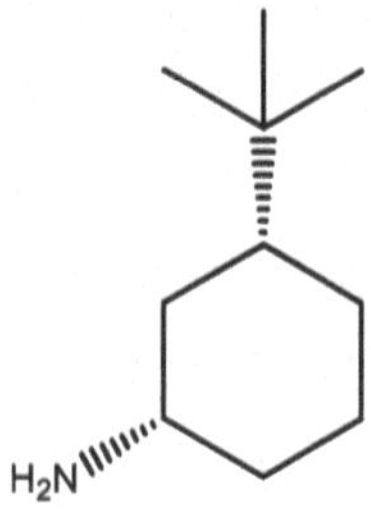

d)

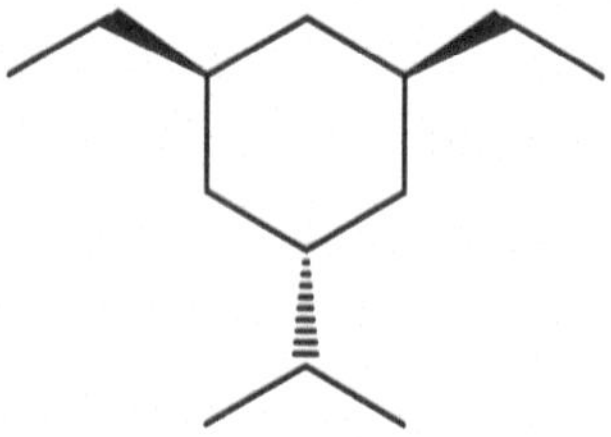

Functional Groups

3

What you will learn:

- Importance of knowing functional groups
- Names of most common functional groups
- Identification of functional groups

3.1 INTRODUCTION

One of the knowledge pillars in organic chemistry is the ability to recognize and distinguish between each chemical functional group. A simplified view of a functional group would be to think of a functional group as "where the action happens" or where the chemistry takes place. It is important to note that while we are investigating common functional groups, it is by no means a complete list; in fact, much of organic chemistry can be thought of as a series of functional group interconversions. This section is set up in groups of related functional groups to make these groups easier to get to know. It is important that when looking at functional groups we consider all the atoms involved in the functional group, the change of one atom can result in a new functional group. You will see examples for each functional group correctly identifying the atoms involved. If this idea of organic chemistry as a series of functional group interconversions is considered, then the fundamental skill of recognizing functional groups can be one of the most important skills to set you up for success.

3.2 LIST OF MOST COMMON FUNCTIONAL GROUPS

- Alkane
- Alkene
- Alkyne
- Arene
- Halide
- Alcohol
- Phenol
- Amine
- Nitro
- Nitrile
- Thiol
- Ether
- Sulfide (thio ether)
- Disulfide
- Aldehyde

- Ketone
- Carboxylic acid
- Ester
- Amide
- Acid halide (acyl halide)
- Carboxylic acid anhydride (anhydrides)

3.3 FUNCTIONAL GROUPS—ALKANE, ALKENE, ALKYNE, AND ARENE

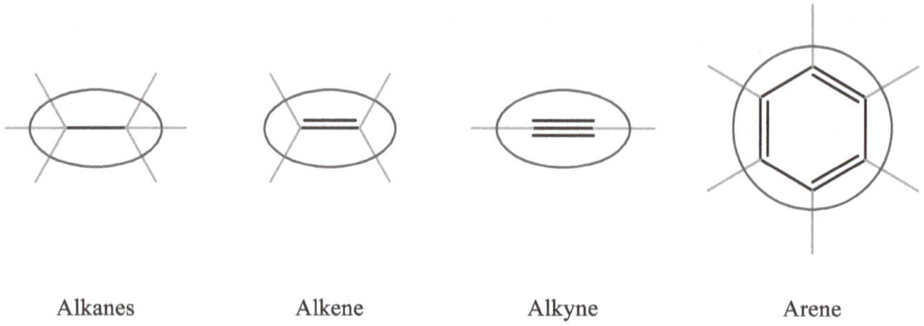

Figure 3-1: ***Hydrocarbon-based functional groups highlighting the correct atoms.***

These are the hydrocarbon-based functional groups, meaning they include only hydrogens and carbons in the functional moiety (Figure 3-1). These are the simplest groups; in particular, alkanes are the backbone of most organic molecules and are so common that we often don't talk about alkanes as functional groups. As you can easily see from the name, alkane, alkene, and alkyne only vary by one letter, nothing can be confusing about that; while the change in name is small, the reactivity and behavior is large. Alkanes are completely saturated; looking back to general chemistry, they employ all single bonds and sp^3 hybridization with tetrahedral geometry; these are often referred to as alkyl chains. Alkenes, also called double bonds, now introduce a single unit of unsaturation between two atoms; this is represented by a pi bond between the two atoms and loss of a sigma bonding orbital on each rehybridizing to sp^2 centers and trigonal planar spatial arrangement. Alkynes introduce two units of unsaturation between two atoms; this is illustrated by two pi bonds between the atoms and loss of two sigma bonding orbitals resulting in a triple bond with sp hybridization and linear geometry. When identifying alkenes and alkynes, you only need to include the atoms making the unsaturation; another way to look at this is just the atoms involved in the double or triple bonds respectively. Arenes are the most complex of the hydrocarbon functional groups. These are moieties that are six-membered rings with alternating units of unsaturation, most often we call these by their more colloquial names—benzene rings and are often referred to as aryl (Figure 3-2).

Practice

3.3.1. Circle and name the functional groups in the molecules below. Make sure you circle all necessary atoms to identify each functional group. (Do NOT circle alkanes)

a)

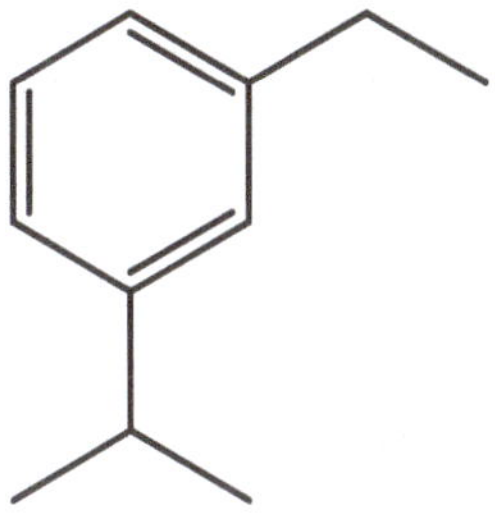

b)

c)

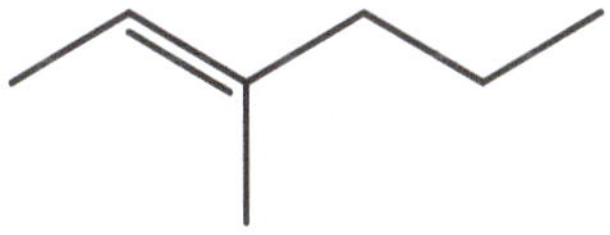

d)

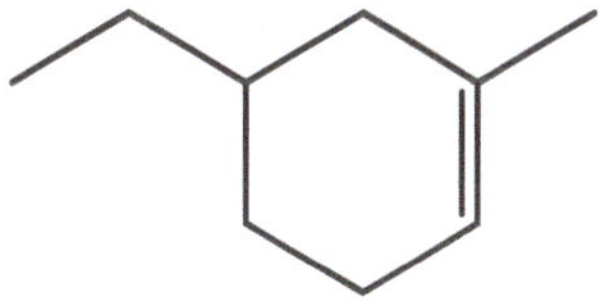

e)

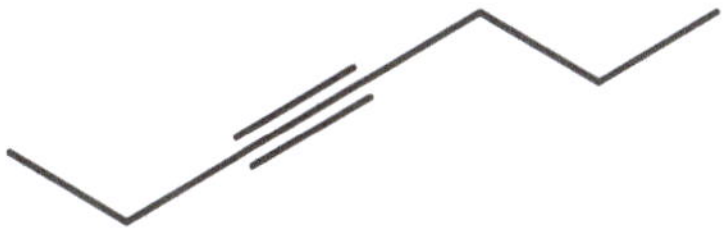

f)

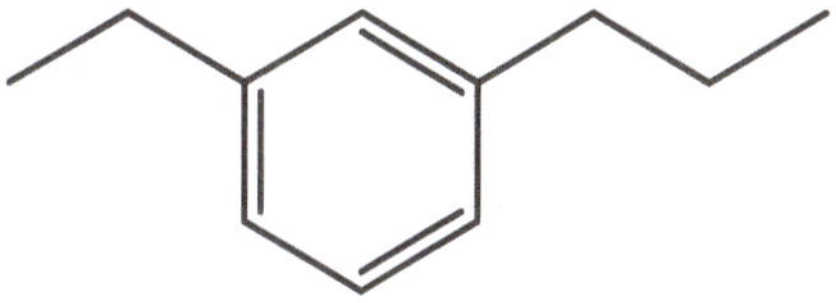

3.4 FUNCTIONAL GROUPS—HALIDE, ALCOHOL, PHENOL, AMINE, NITRO, NITRILE, AND THIOL

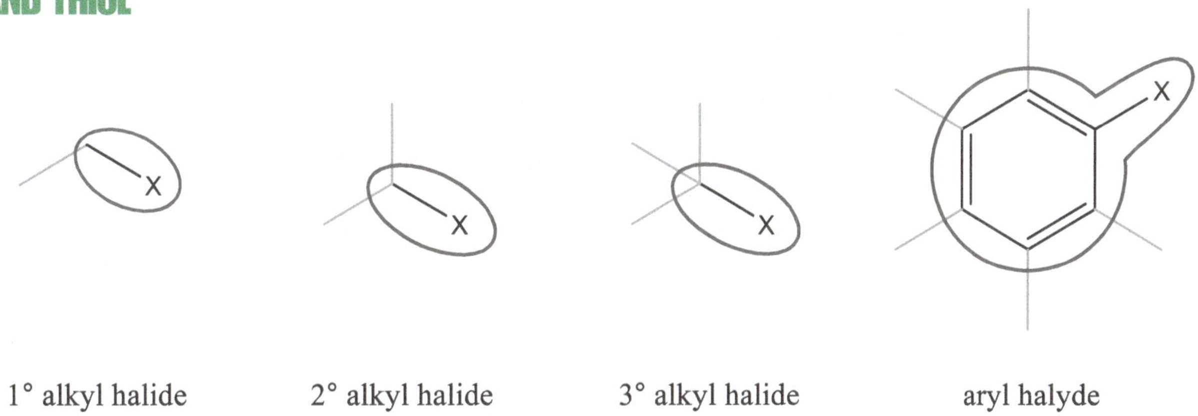

Figure 3-2: Alkyl halides being shown as 1°, 2°, 3° and aryl variants.

Now we want to add some heteroatoms, in the previous section we mentioned that the alkanes are the backbone to most organic molecules; we will now add some chemical groups to those backbones. **It is important to note that in this section we are not just looking at the added group; it must also include the carbon that the groups are attached to.** Halides are the simplest addition; these are the addition of a halide. If the group is added to an alkane, it is termed an alkyl halide, if it is added to an arene it is called an aryl halide.

Alcohols are the attachment of a hydroxyl group (-OH) to a carbon; phenols are when the hydroxyl is attached to a carbon on an arene ring (Figure 3-3). The carbons that are part of the halide and alcohol functional groups are important because they define the type of halide or alcohol and impact the reactivity; we will talk about this more in Chapter 4 where nomenclature is discussed.

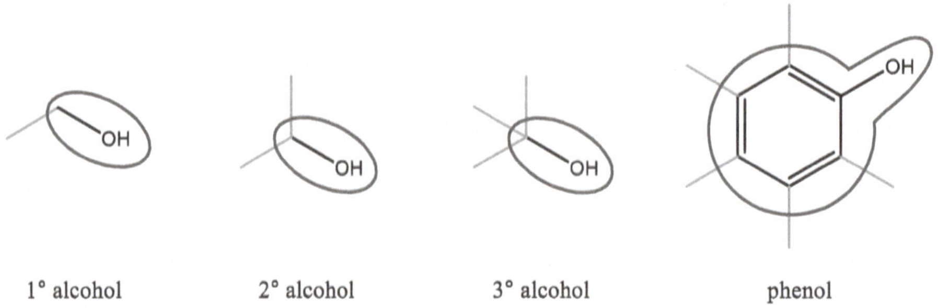

Figure 3-3: ***Examples of 1°, 2°, and 3° alcohols and the aromatic phenol functional group.***

The amine functional group includes nitrogen that is attached to a carbon on an alkyl chain(s). These groups are different from halides and alcohols in their attachment may be up to four different alkyl chains. This attachment pattern will affect the name and reactivity of the amines, which will be discussed in the nomenclature chapter as well (Figure 3-4).

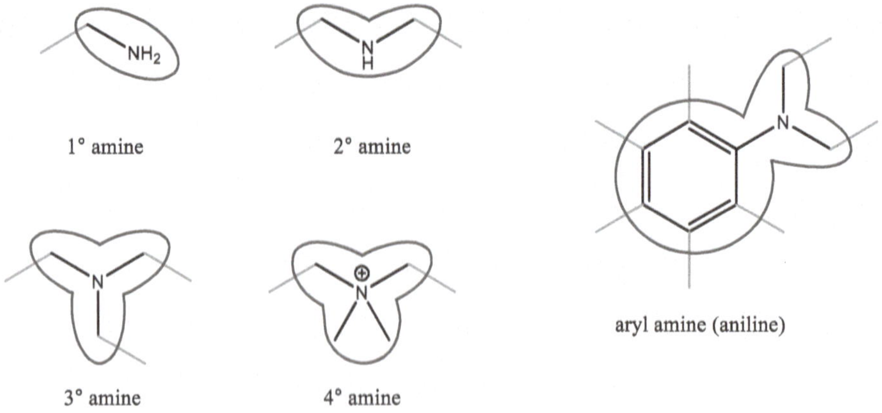

Figure 3-4: ***Examples of 1°, 2°, 3°, and 4° amines. Aryl amines, also known as anilines, are another important class of molecules.***

The nitro functional group contains a nitro (-NO_2) attached to a carbon, this chemistry is seen in organic chemistry when discussing electrophilic aromatic substitution. Nitriles are interesting functional groups; they contain the cyano moiety (-CN triply bonded) attached to a carbon, which occur in both alkyl carbons and aryl carbons. The final group in this section is thiols, which are like alcohols, except the oxygen is replaced with a sulfur atom, which connects to a carbon; these also can occur on alkyl or aryl systems (Figure 3-5).

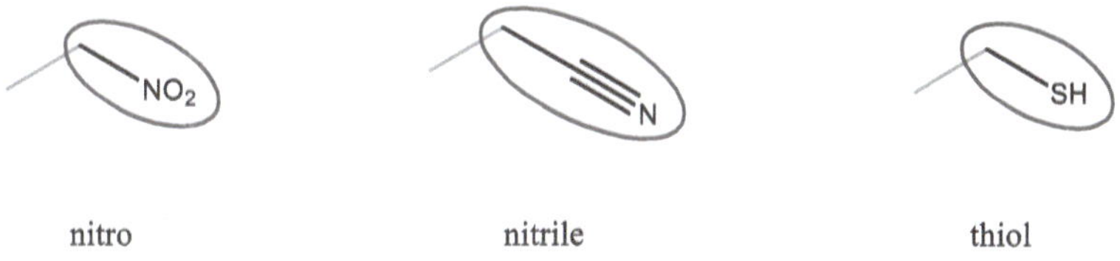

Figure 3-5: ***Other important alkyl-based functional groups with single attachment to the main carbon chain.***

Practice

3.4.1. Circle and name the functional groups in the molecules below. Make sure you circle all necessary atoms to identify each functional group. (Do NOT circle alkanes)

a)

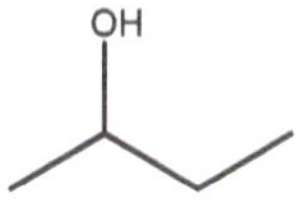

b)

c)

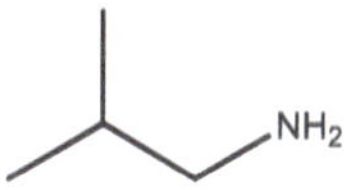

d)

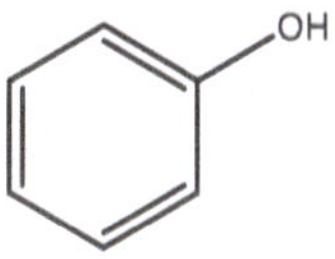

e)

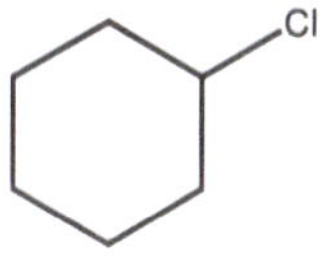

f)

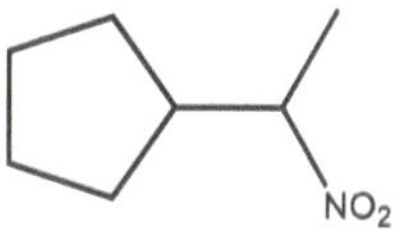

g)

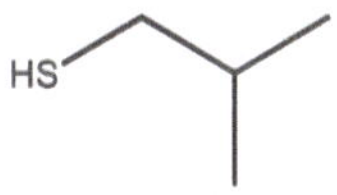

h)

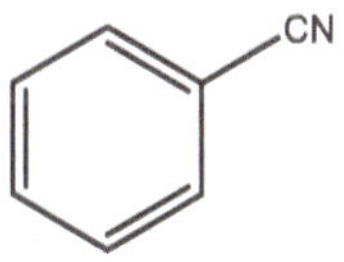

i)

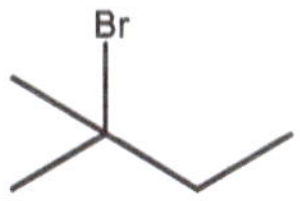

j)

k)

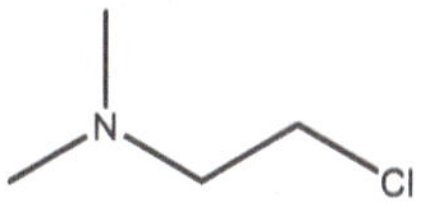

l)

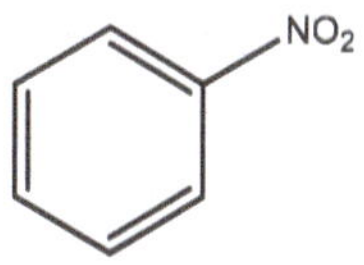

3.5 FUNCTIONAL GROUPS—ETHER, SULFIDE (THIO ETHER), AND DISULFIDE

These functional groups are together because they have the same general look (Figure 3-6).

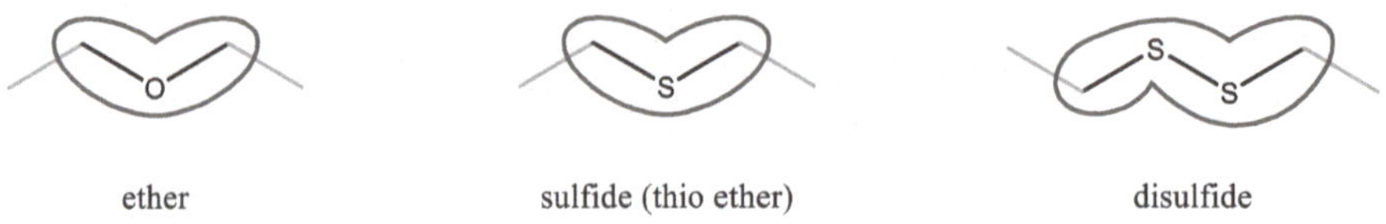

Figure 3-6: ***Oxygen and sulfur containing functional groups with two attachments to carbon.***

Ethers are important chemicals as they are resistant to reactivity and we see them employed as common laboratory solvents. Ethers are oxygens that are attached to two carbon groups, it is important to note that the attached carbons are also included in the functional group. Ethers are often found within cycles as well as acyclic examples. Sulfides look like ethers but the oxygen is replaced with a sulfur atom, hence the other name thio ethers. While these have a similar look to ethers, chemically they are very different. Disulfides look like sulfides but there are two sulfur atoms between the two carbons and have a much different chemical profile than sulfides including a high importance in biochemistry.

Practice

3.5.1. Circle and name the functional groups in the molecules below. Make sure you circle all necessary atoms to identify each functional group. (Do NOT circle alkanes)

a)

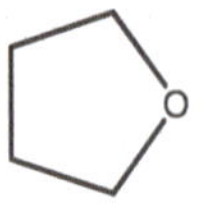

b)

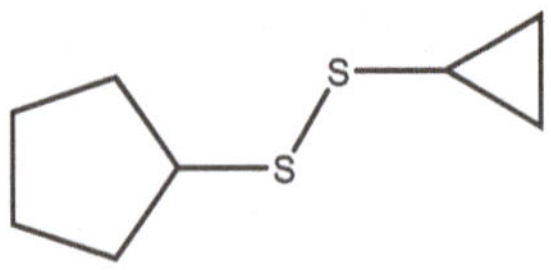

c)

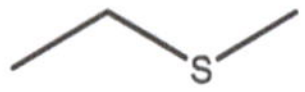

d)

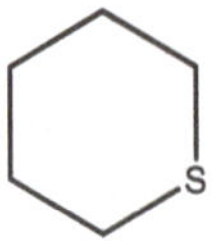

e)

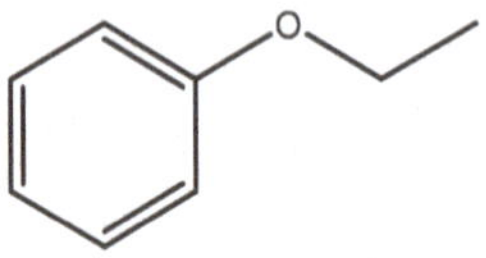

f)

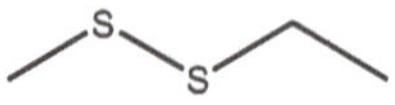

3.6 FUNCTIONAL GROUPS—ALDEHYDE, KETONE, CARBOXYLIC ACID, ESTER, AMIDE, ACID HALIDES (ACYL HALIDES) AND CARBOXYLIC ACID ANHYDRIDE (ANHYDRIDES)

Functional groups in this section all contain the carbonyl moiety; that is a carbon double bonded to oxygen. From our general chemistry knowledge, we know this allows for two other possible bonds to the carbonyl carbon. It is these two bonds that define the different functional groups. A common mistake is simply identifying the carbonyl as a functional group (commonly mistaken for a ketone), carbonyls are not functional groups but part of a functional group. Of the groups we have, a carbon group, alkyl, or aryl, is always one of the possible bonds, the other will vary.

Two very similar looking members of this group are aldehydes and ketones. Aldehydes have a hydrogen as the varying group, in contrast ketones will have two carbon groups attached to the carbonyl carbon. Aldehydes and ketones have a lot of similar chemical characteristics but deviate from each other in reactivity due to their structural differences (carbon vs. hydrogen). The next important class is carboxylic acids, these have a carbon on one side of the carbonyl and a hydroxyl on the other. These maybe the most recognizable functional groups in this section (Figure 3-7).

Figure 3-7: ***Carbonyl containing functional groups.***

The remaining groups are known as the derivatives of carboxylic acids, this is because they can all be synthesized from carboxylic acids. Esters look like carboxylic acid, but they have alkoxy group (-OR) attached to the carbonyl, amides look similar but have an amino group attached ($-NR_2$, where R is carbon or hydrogen). Esters and amides groups are very common in biological systems and therefore you will see them often. Acid or acyl halides look like carboxylic acid, but the hydroxyl has been replaced with either a chloride or bromide (fluorides and iodides are not used). The carboxylic acid anhydride or simply anhydrides have a look of two carboxylic acids joined together through a single oxygen. Acid chlorides and anhydrides are very reactive and are very useful synthetic intermediates (Figure 3-8).

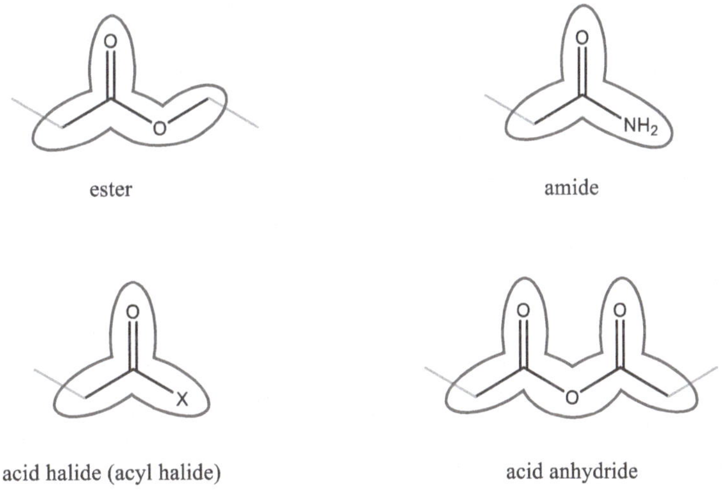

Figure 3-8: ***Carboxylic acid derivative functional groups.***

Practice

3.6.1. Circle and name the functional groups in the molecules below. Make sure you circle all necessary atoms to identify each functional group. (Do NOT circle alkanes)

a)

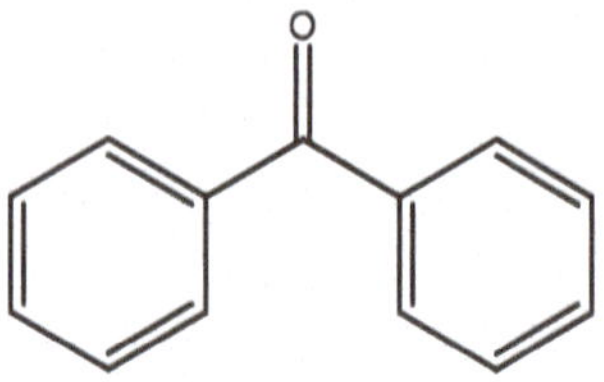

b)

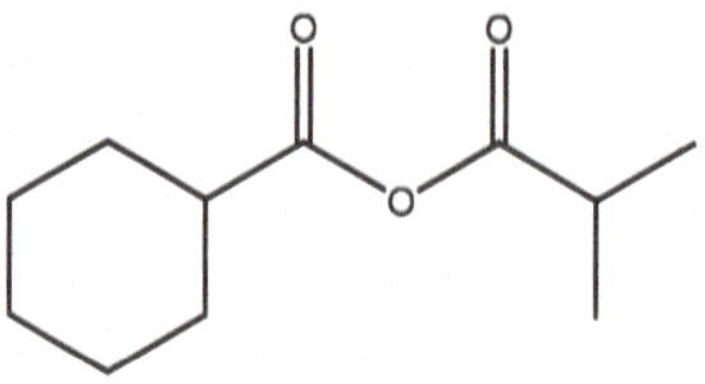

c)

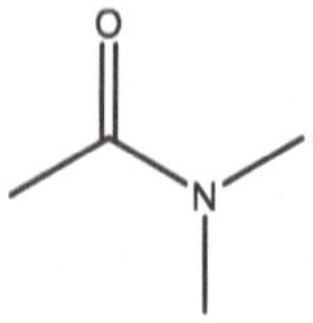

d)

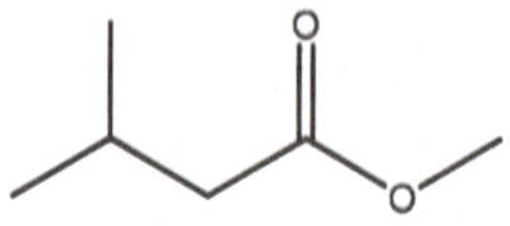

e)

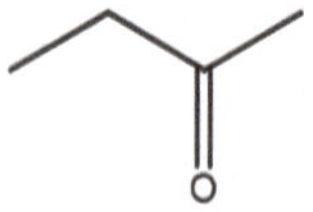

f)

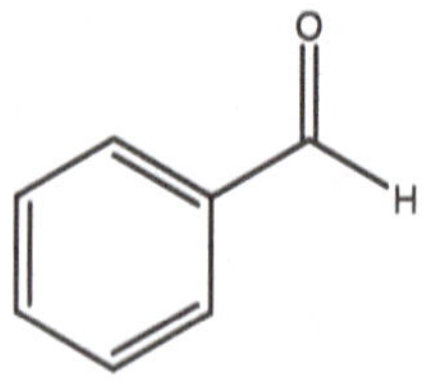

g)

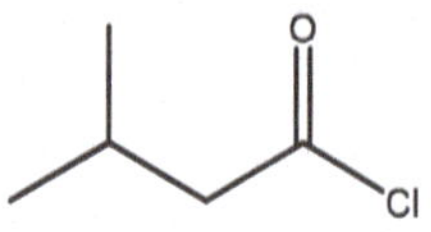

h)

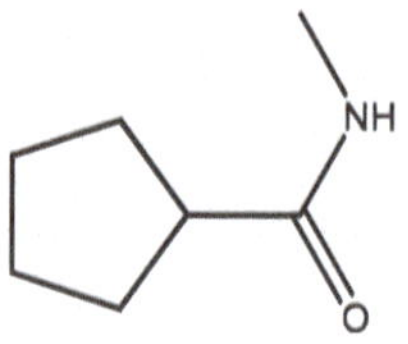

i)

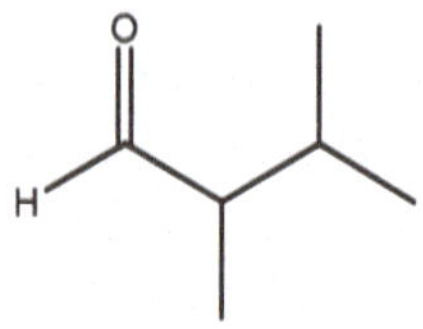

j)

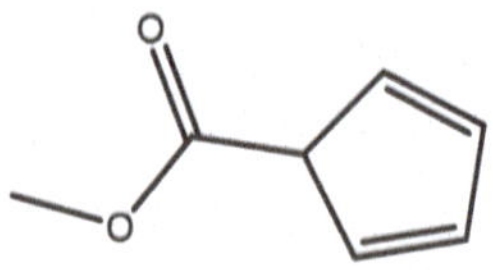

3.7 FUNCTIONAL GROUPS SUMMARY

In summary you will encounter several other functional groups, often they will be combinations of the above groups such as enol or enynes, combinations of an alkene and alcohol and alkenes and alkynes respectively. This could be complex like the enol, having the alcohol on the alkene or they can be like the enyne where the alkene and the alkyne are in series with each other. The most important skill is to practice these and get good at identifying the functional groups as it will make your time in organic chemistry much more productive and the material will be easier to master.

Name: ____ Date: ____________

INTEGRATED PROBLEMS

3.1. Name the functional groups in the molecules below.

a)

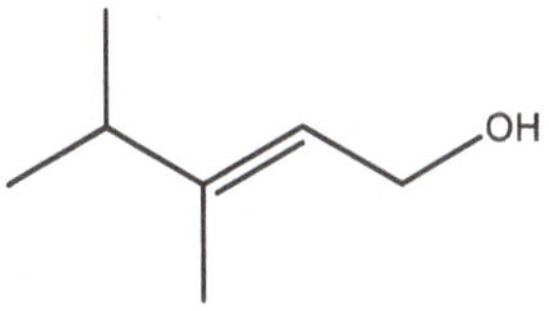

b)

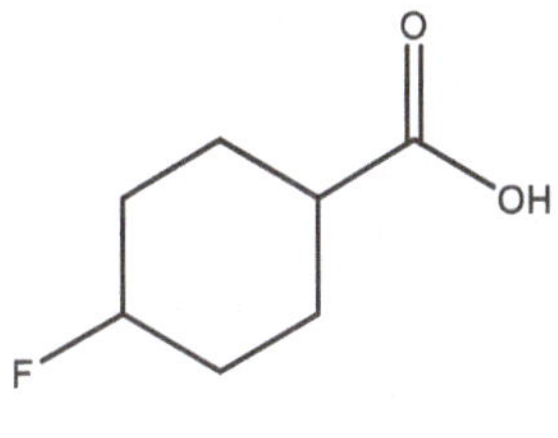

c)

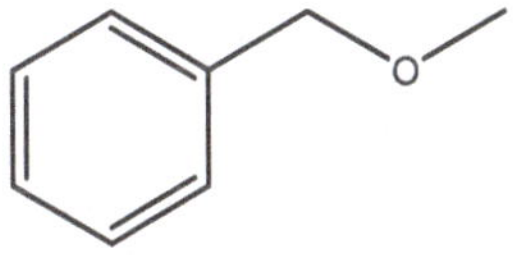

d)

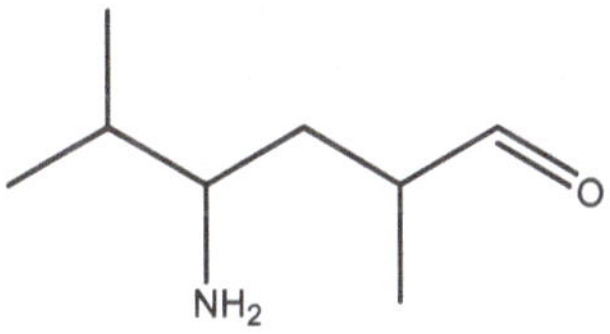

e)

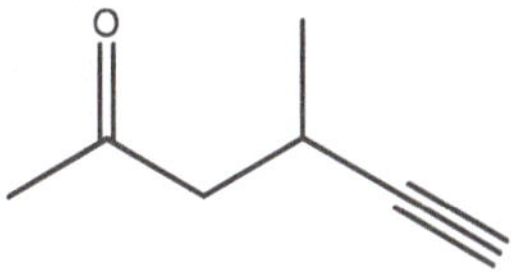

f)

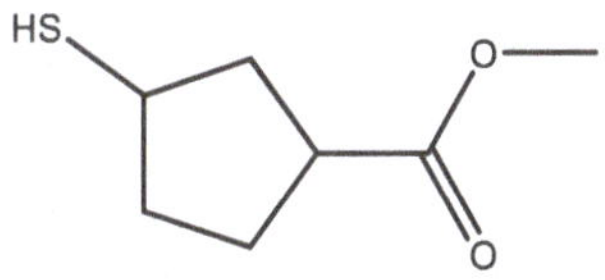

g)

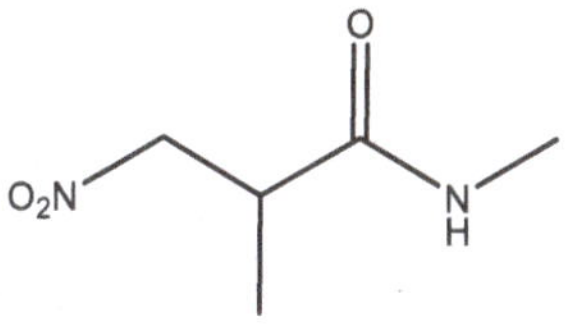

h)

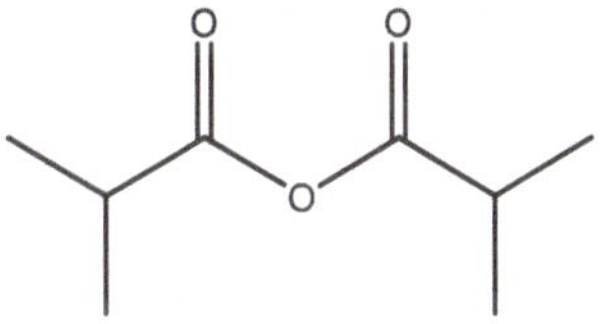

i)

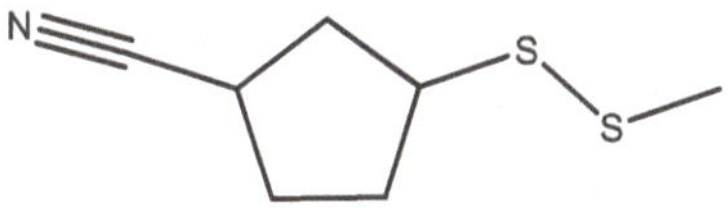

j)

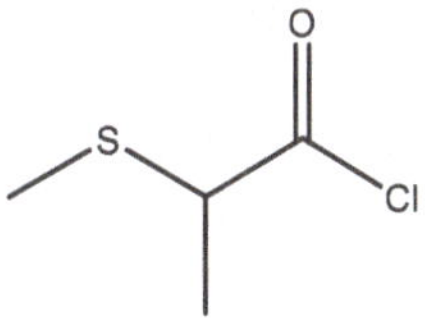

3.2. Draw the functional groups for the names below. Use "R" as the alkyl group where necessary.

a. Aldehyde

b. Ester

c. Nitrile

d. Carboxylic acid

e. Arene

f. Disulfide

g. Alcohol

h. Ether

i. Thiol

j. Ketone

k. Halide

l. Amine

3.3. Circle and name the functional groups in the molecules below. Make sure you circle all necessary atoms to identify each functional group.

a)

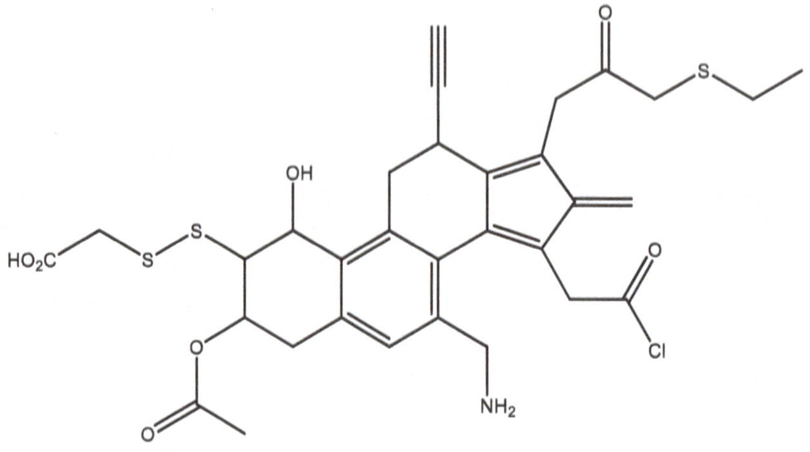

b)

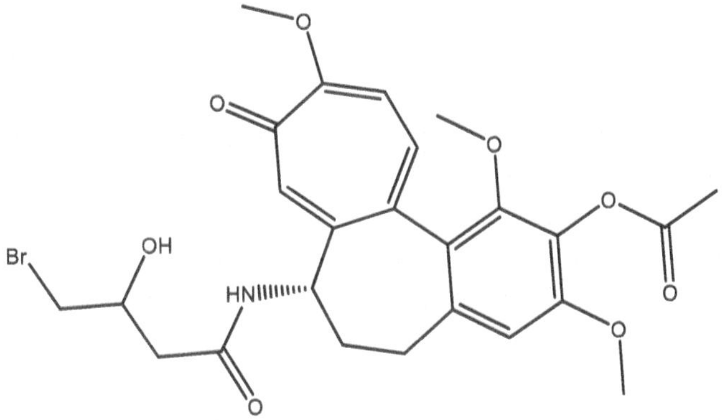

c)

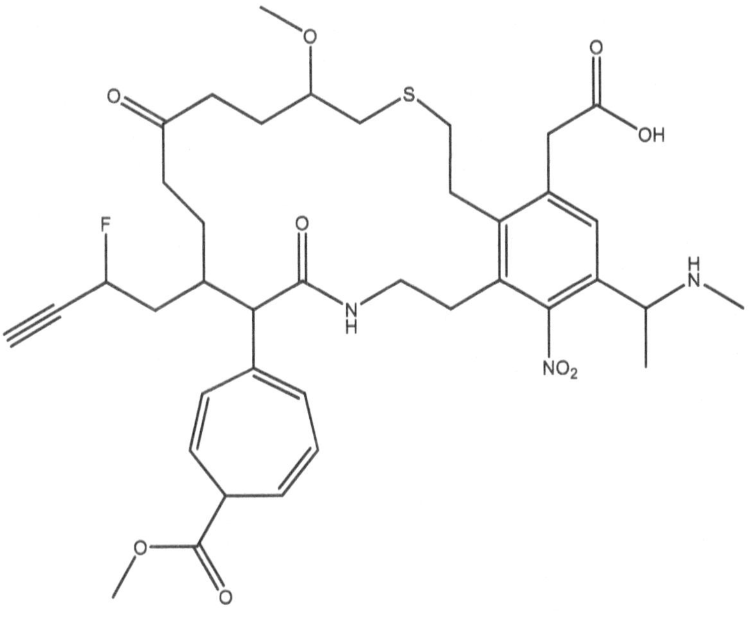

Naming Organic Molecules

4

What you will learn:

- IUPAC naming convention
  - Adding functional groups
  - Adding stereochemistry to names
  - Complex branch naming
- Importance of nomenclature
- Common names

4.1 INTRODUCTION

So why do we have to go through learning a complex way to name molecules, the answer is actually very simple ... to make your life easier. Easier is not typically what comes to mind to an organic chemistry student when talking about naming chemical compounds; however, what if you wanted to talk to a friend in a different school/state/country about a chemical you were learning about in your course. It would be important to be able to clearly communicate to them in a very precise way so that they could visualize the same chemical exactly as you are communicating it to them. Having a methodology to communicate in a very precise fashion is fundamentally important to all areas of sciences; this is especially true in chemistry where it is possible to have a potentially hazard chemical reaction occur if the chemical is not interpreted correctly.

Let's consider what is going on in the naming convention. The naming convention we use in organic chemistry was developed by the International Union of Pure and Applied Chemistry (IUPAC) and had been codified into the IUPAC Compendium on Chemical Terminology (colloquially referred to as the gold book). Why is this group making determinations on how we name chemicals? Well, IUPAC is the governing body that sets standards in the field of chemistry. The idea behind the naming convention is to unambiguously define each chemical distinctly from another. While this sounds like a lot of work, correctly naming compounds can help us in our determination of isomers and several other properties (Figure 4-1). It is important to note that we will not be learning the entire gold book (it is a vast resource covering all aspects of chemistry most of which are more advanced than this course), but just enough to allow us to correctly communicate the chemicals of interest.

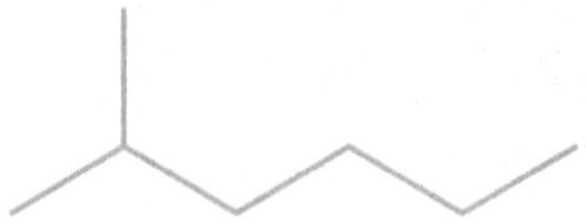

2-methylhexane

Locant	prefix	parant	suffix
location	substituent	main chain	functional group

Figure 4-1: ***Basic components of naming.***

4.2 IMPORTANT DEFINITIONS

- **Main/parent chain:** Longest continuous chain of carbons
- **Branches/substituents:** Atom or group of atoms which replaces one or more hydrogen atoms on the parent chain
- **Alkyl groups:** Substituent formed from alkanes by removing one hydrogen
- **Common names:** Well established and well-known names for compounds commonly used in industry
- **Complex substituents:** Branch substituent without a common name, therefore the systematic naming rules must be used
- **Functional groups:** Atom or groups of atoms that are the reactive part of the molecule

Basic Naming Rules

STEP 1 - Identify the parent chain

A. Find the longest continuous chain of carbons. Use this name as the parent name of the compound.

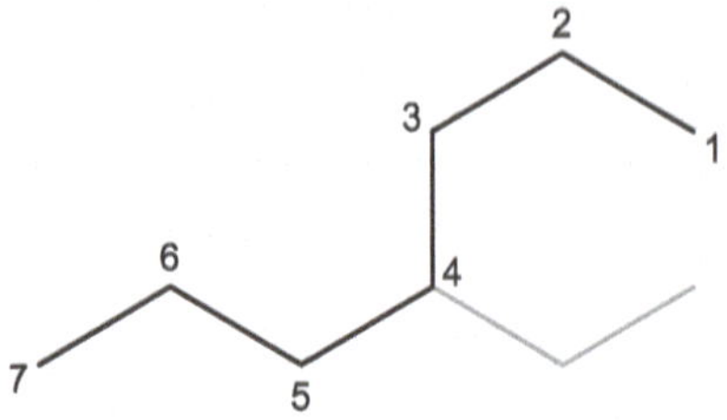

7 carbon: heptane

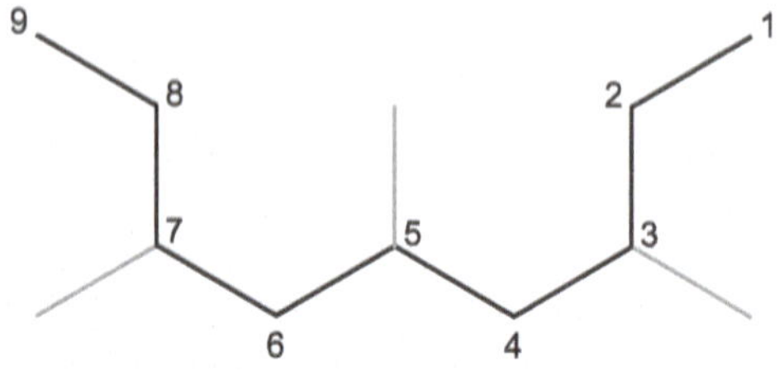

9 carbon: nonane

B. If there are two different chains of equal length: choose the chain with the larger number of branches (substituents)

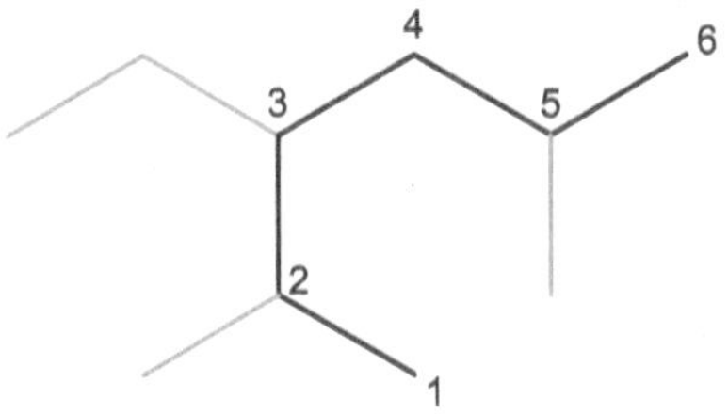

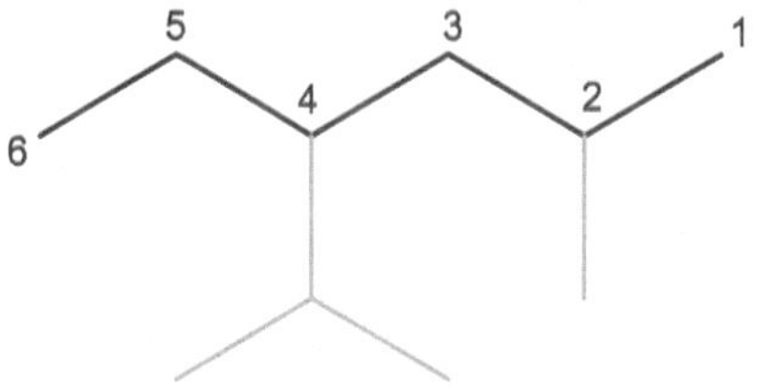

3 substituents/branches

CORRECT

2 substituents/branches

INCORRECT

STEP 2 - Number the atoms in the parent chain

A. Begin numbering at the end nearest to the first branching point

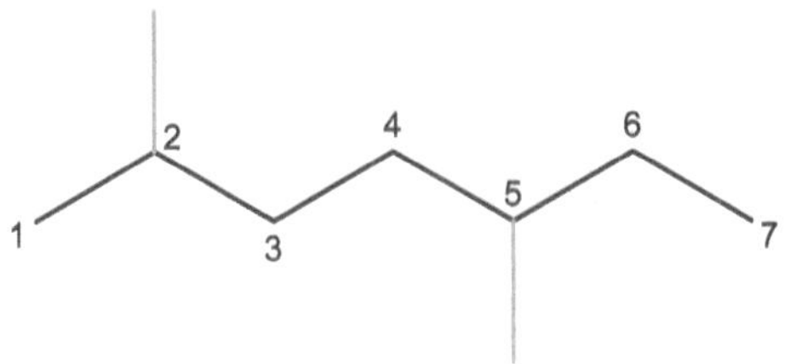

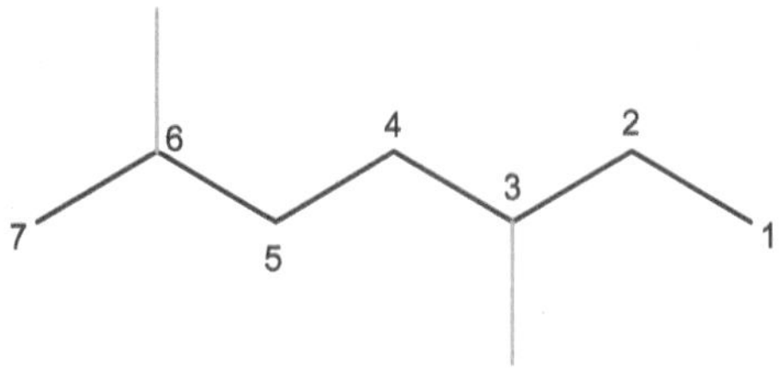

1st substituent on carbon 2

CORRECT

1st substituent on carbon 3

INCORRECT

B. If there is branching at equal distance from either end, base the numbering on the next branching point

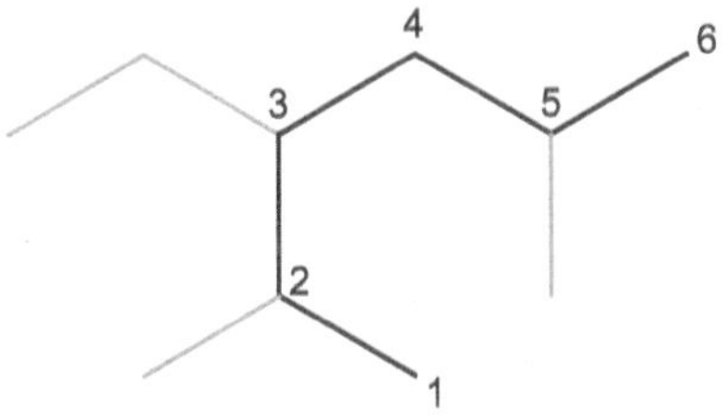

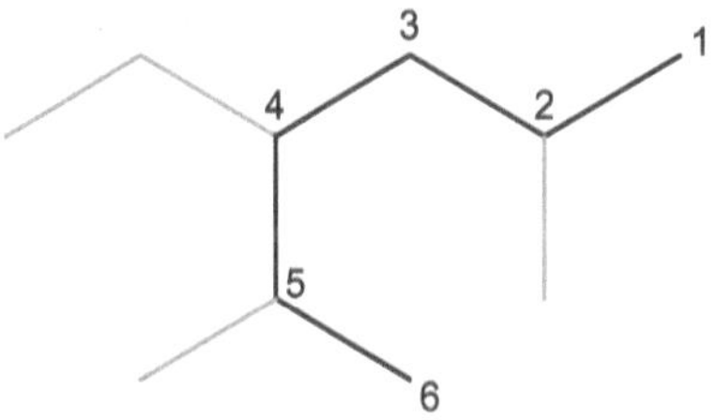

2nd substituent on carbon 3

CORRECT

2nd substituent on carbon 4

INCORRECT

STEP 3 - Identify and number the substituents

A. Assign a number to each substituent according to its point of attachment on the carbon chain

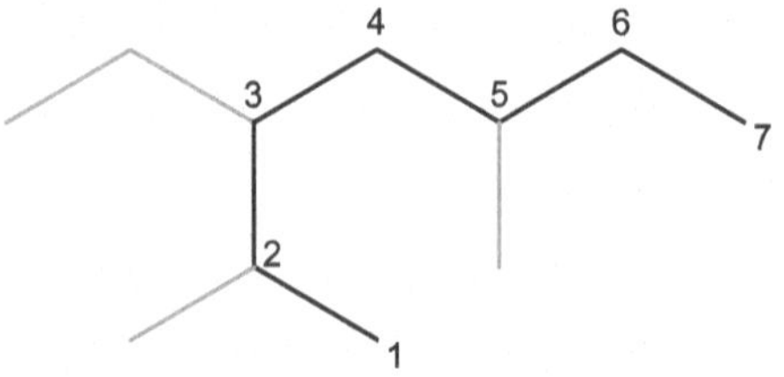

list of substituents with their numbers:
2-methyl
3-ethyl
5-methyl

B. If two substituents are on the same carbon, give them the same number

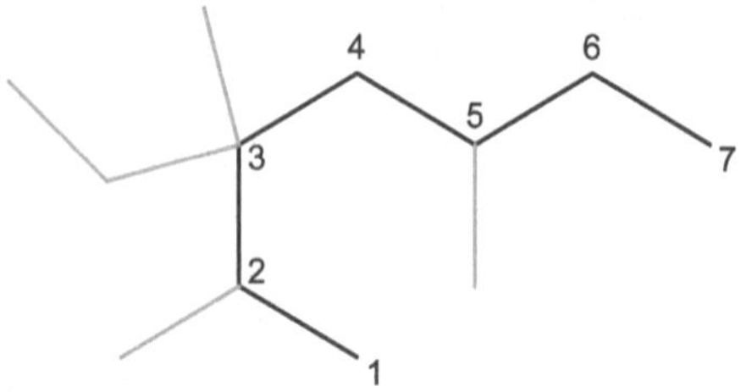

list of substituents with their numbers:
2-methyl
3-ethyl
3-methyl
5-methyl

STEP 4 - Write the name as a single word

Use hyphens to separate different prefixes
Commas separate numbers
Two or more substituents - cite them in alphabetical order
Two or more identical substituents - use multiplier prefixes (di, tri, tetra, penta, hexa, etc....)[multiplier prefixes are not used for alphabetizing]

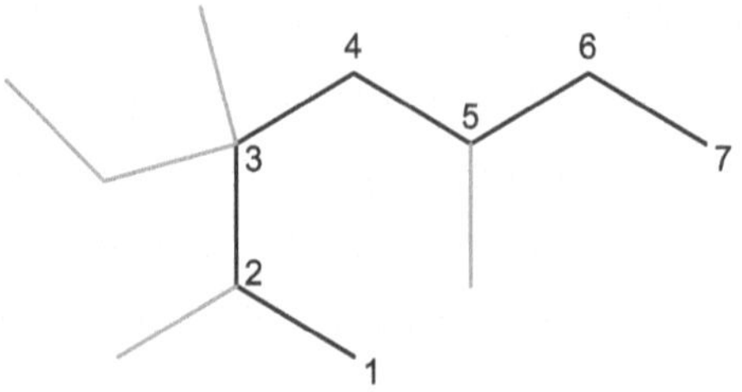

3-ethyl-2,3,5-trimethylheptane

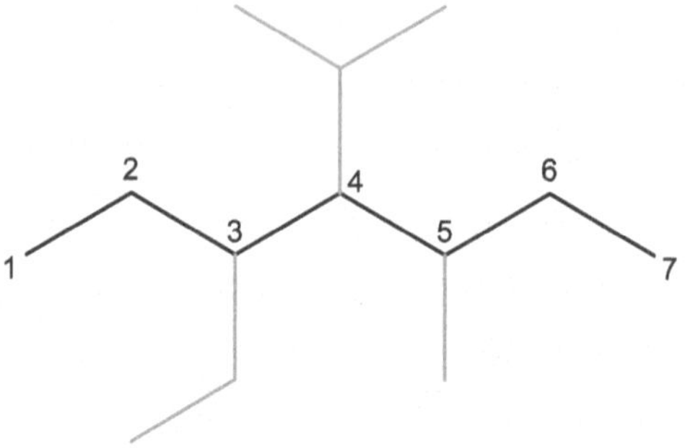

3-ethyl-4-isopropyl-5-methylheptane

STEP 5 - Name complex substituents as though it was itself a compound

Begin numbering the complex substituent at the point of attachment to the main chain

The substituent is alphabetized using the first letter in its complete name and is set off in parentheses

Iso and neo prefixes are used for alphabetizing, but sec and tert are not

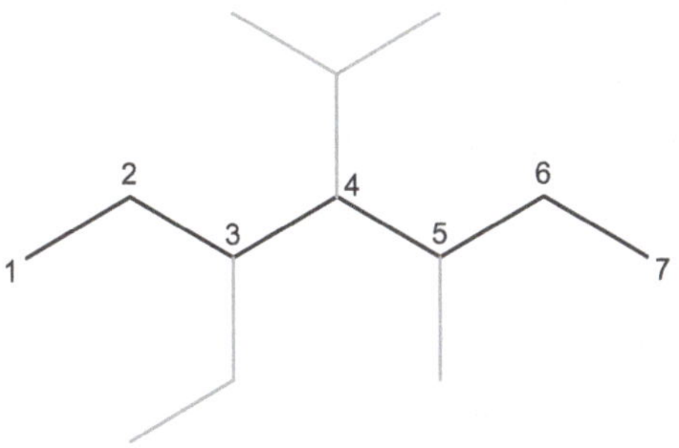

3-ethyl-4-isopropyl-5-methylheptane

or

3-ethyl-5-methyl-4-(1-methylethyl)heptane

Practice

4.2.1. Provide the IUPAC names of the following alkanes.

a)

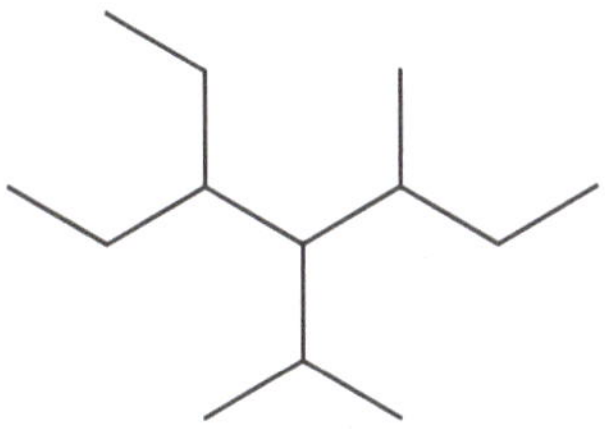

b)

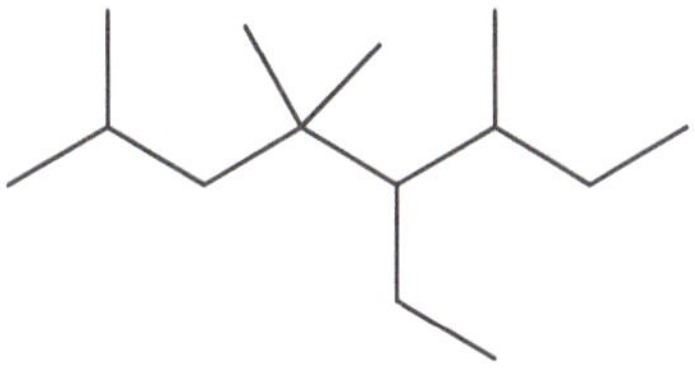

c)

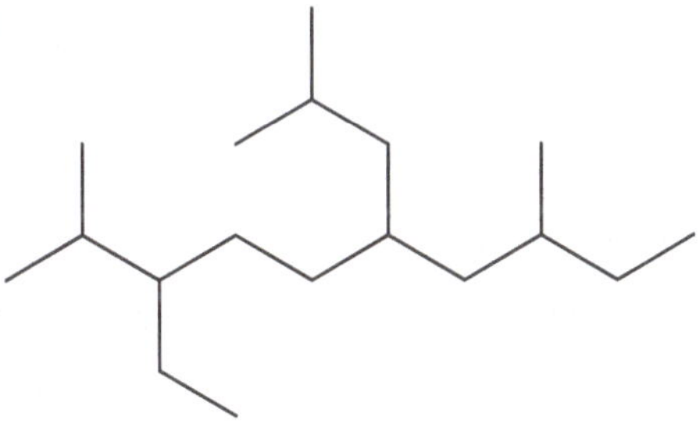

d)

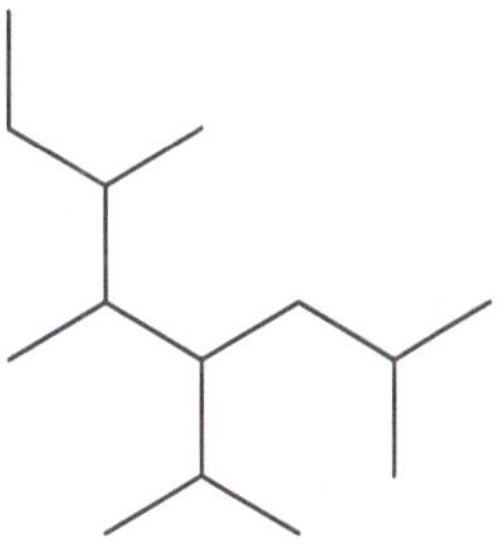

e)

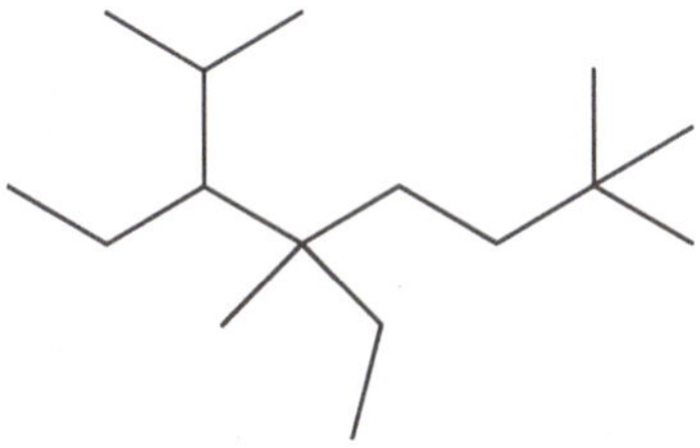

f)

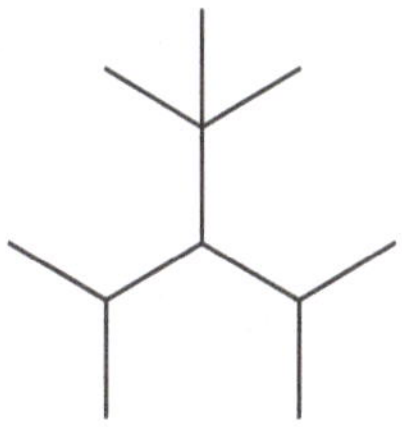

4.2.2. Draw to structures corresponding to the following names.

a. 4-Ethyl-2,4,7-trimethyloctane

b. 3,4-Diethyl-2,2,3,5-tetramethylhexane

c. 2,2-Dimethylheptane

d. 8-*tert*-Butyl-3,11-diethyl-5-propyltridecane

e. 4-Ethyl-5-isobutyl-6-isopropyl-2,2-dimethyldecane

4.3 ADVANCED NAMING RULES

Other Functional Groups

These basic naming rules seem easy enough and the basics are what students typically have a difficult time mastering; however, what do we do when our compound contains other functional groups other than alkanes? The answer is actually really simple: we go back to the basics. We still use the basic rules, but we need to now change what the parent chain is. The parent chain will always contain the highest priority functional group or in some cases such as enynes (alkenes and alkynes) or enones (alkenes and ketones), the parent chain must contain both functional groups. In the following table you will find a list of functional groups, their suffix ending as a parent and their name when they are a substituent.

FUNCTIONAL GROUP	PARENT CHAIN NAME	SUBSTITUENT NAME
Alkane	-ane	-yl
Alkene	-ene	-enyl
Alkyne	-yne	-ynyl
Alcohol	-ol	-hydroxy
Amine	-amine	-amino
Thiol	-thiol	-mercapto
Ether	Ether	-oxy
Aldehyde	-al	-carbaldehyde
Ketone	-one	-oxo
Carboxylic acid	-oic acid	-carboxylic acid
Ester	-oate	
Amide	-amide	
Nitrile	-onitrile	-carbonitrile

Table 4-1: ***A list of common functional groups and their names as used in the IUPAC naming convention. These functional groups are listed in increasing priority; it is important to note that this is not a complete list of all functional groups.***

*Halides in compounds are given the same priority as alkanes; there are not parent chains; however, as a substituent, the -ine ending is replaced with -o.

4.4 CYCLIC MOLECULES

As we saw above the standard naming convention works well because it is a rigorous, systematic approach to naming a molecule. All the examples in the preceding section dealt with linear or acyclic molecules, can the naming convention handle cyclic molecules? The answer is yes it can, we will simply add the prefix of cyclo- to the new parent chain that is the cyclic portion of the molecule. What if there is a both a cyclic and an acyclic portion of the molecule? Yes, the convention still works; however, we first must meet certain criteria. Firstly, we choose the parent based on the functional group, if the functional group is on the cycle, the cycle is the parent. If the functional group is on the acyclic portion of the molecule, the acyclic portion is the parent. The next criteria we encounter is what to do if they have the same level of functional group priority. The second criterion is the longer portion becomes the parent and the shorter portion is the substituent, while if they are the same size, the cycle becomes the parent. Something important to note, a common mistake that students make is counting the carbon on the ring for both the acyclic and cyclic chain (Figure 4-2).

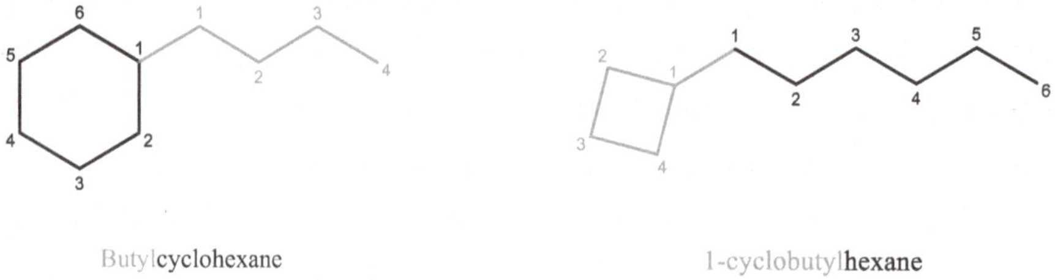

Figure 4-2: ***Naming cycles.***

Benzene Rings and Common Names

It is said earlier in this chapter, the reason this works is it is a robust systematic process to name molecules, with that being said, you will still find non-IUPAC naming. Many chemicals have well known, well established names that are used in industry and academics, this is especially important in chemistry labs (even chemical manufactures will have the common name as the preferred name on the reagent bottle). For instance, most people refer to 2-propanol by its common name of isopropanol or methylbenzene as toluene; this is where it is important to know some common names to help decipher. A common area where naming gets very confusing is when naming bezene rings as these common names are prevalent in the naming of the parent chain.

Stereochemistry

The last important caveat in systematic naming is the placement of stereochemistry (a concept that occurs in a later chapter). The absolute and relative stereochemistry is an important distinction between our stereoisomers. Firstly, we will look at the relative stereochemistry. Relative stereochemistry such as *cis*- is written before the name in italics and followed by a hyphen (Figure 4-3). Absoluter stereochemistry is similar in that it comes before the name and has a hyphen separating; however, absolute stereochemistry such as R or Z must be placed inside parentheses (Figure 4-3). It is important to note that if there is more than one type of absolute stereochemistry occurring in the molecule, you must include the location. One of the most commonly missed naming procedures is failure to include the appropriate stereochemistry in the name.

Figure 4-3: ***Examples of including stereochemistry in the systematic naming scheme.***

Practice

4.4.1. Provide the IUPAC names of the following molecules

a)

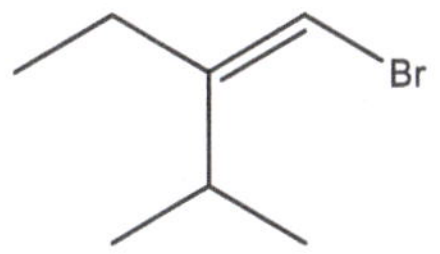

b)

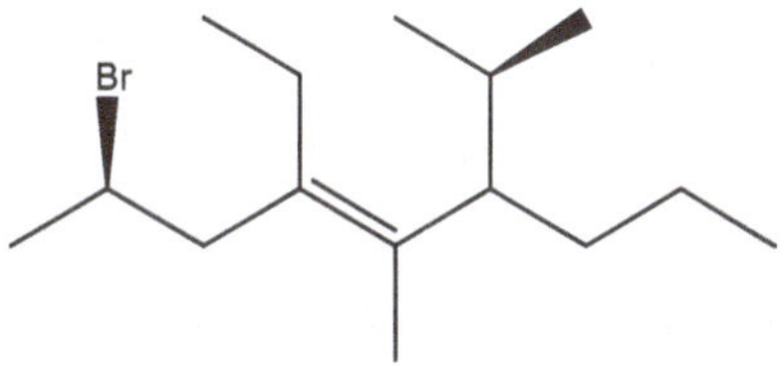

c)

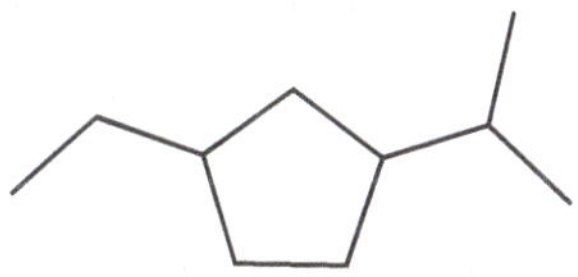

d)

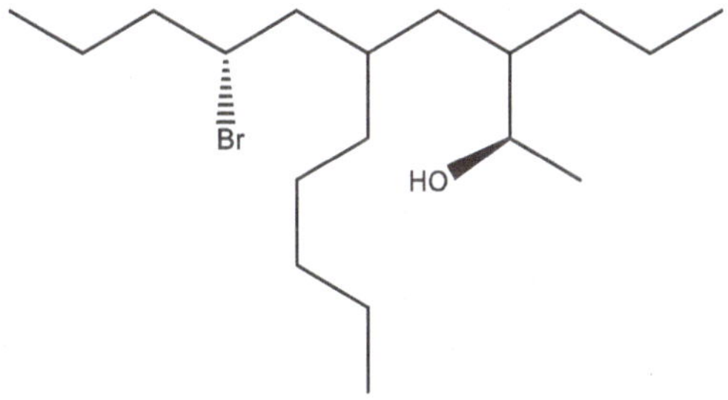

e)

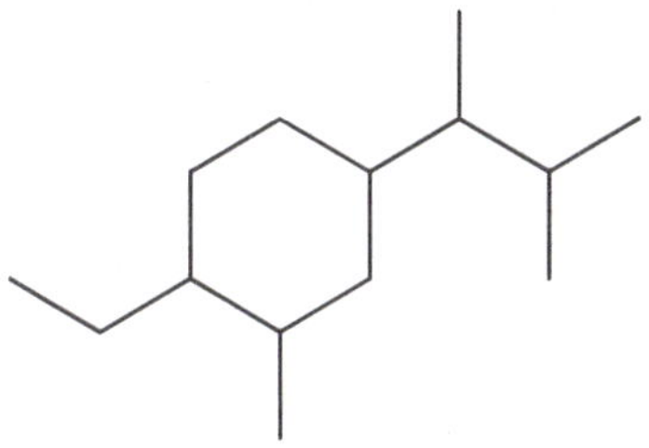

f)

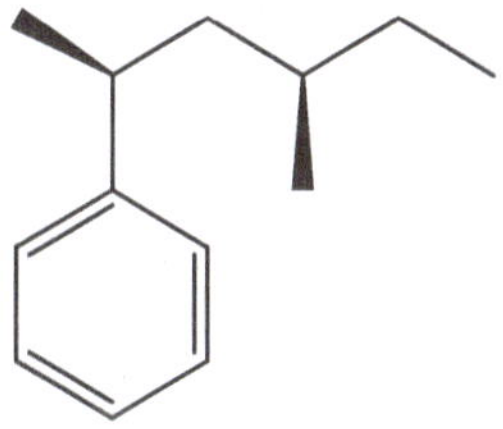

g)

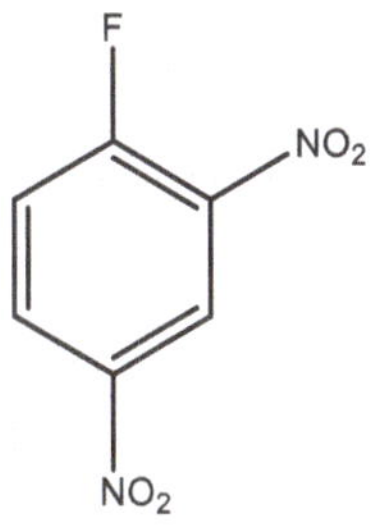

h)

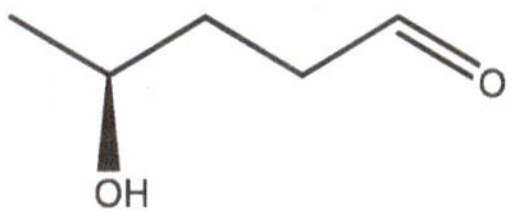

4.4.2. Draw the structures corresponding to the following names

a. *R*-2-bromo-3-hexanone

b. *N*,*N*,4-trimethylpentanamide

c. 1-Methyl-4-(methylethyl)cyclohexane

d. 3-Aminohexane

e. 3-Chlorobenzaldehyde

4.5 SUMMARY

You will see naming as you go through your organic chemistry course and your time spent studying the sciences. It is an important part of how we communicate the chemicals we use not only in the chemistry lab, but in several other disciplines, often referring to the fatty acid, the alcohol, or other functional group to talk about molecules. Having a good grasp on how naming works will help you throughout your careers. Once you master the systematic naming convention you will have mastered one of the fundamental pillars of organic chemistry!

Name: ____ Date: ____________

INTEGRATED PROBLEMS

4.1. Provide the IUPAC names of the following molecules

a)

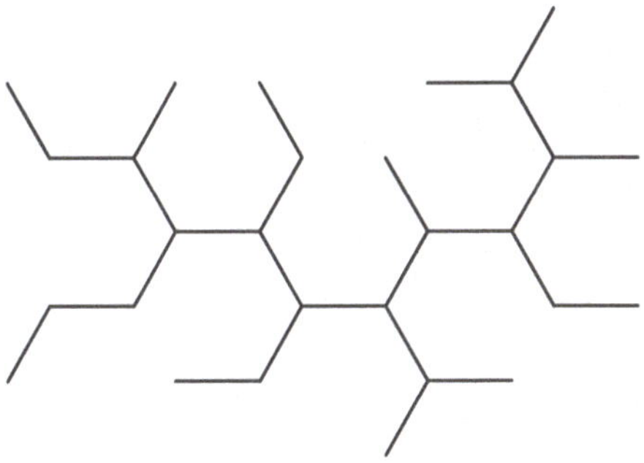

b)

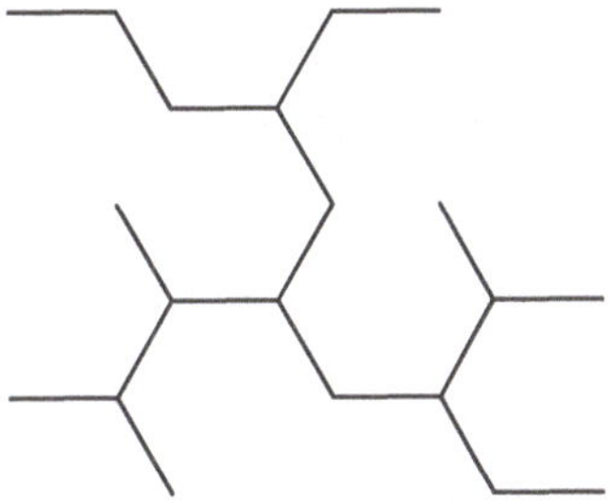

c)

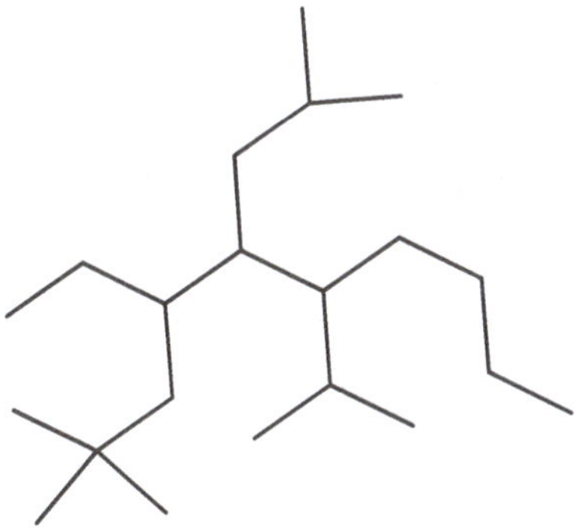

d)

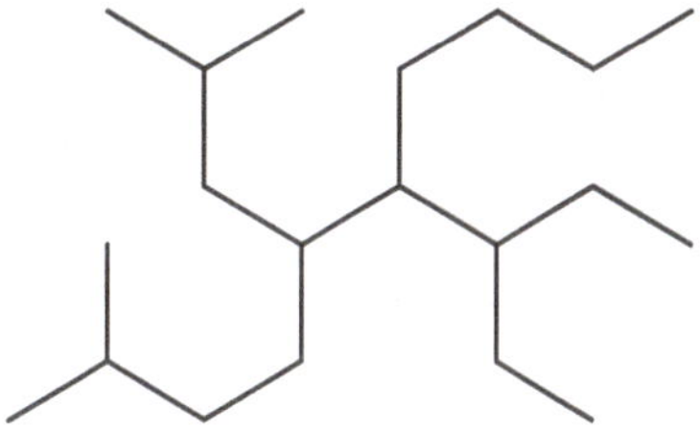

e)

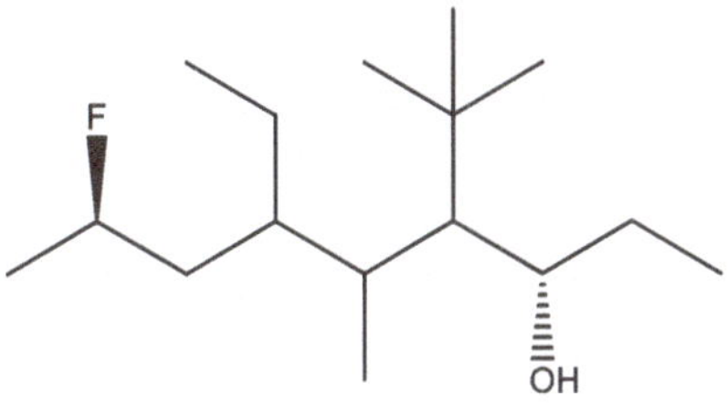

f)

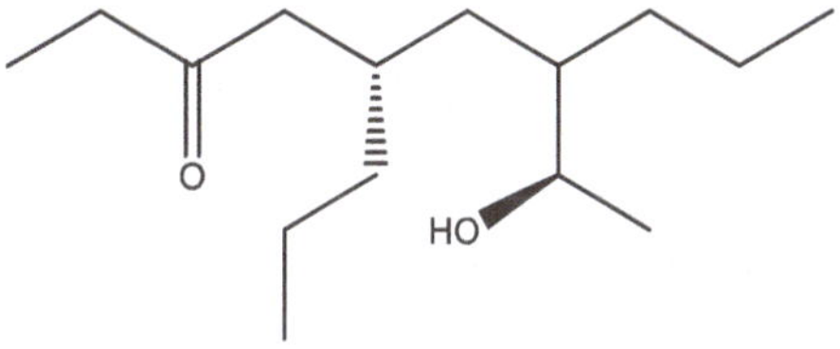

g)

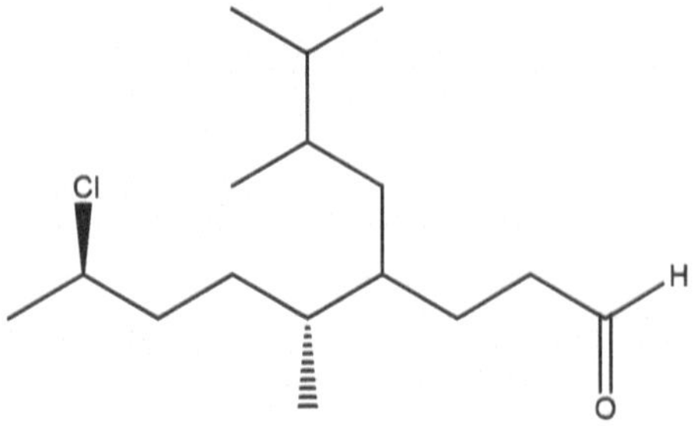
Cl
H
O

h)

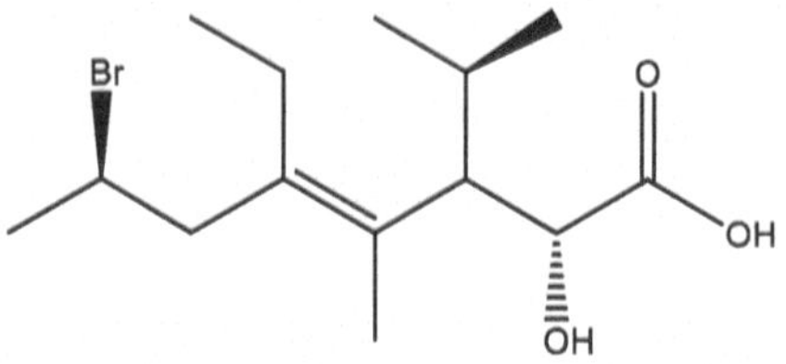
Br
O
OH
OH

i)

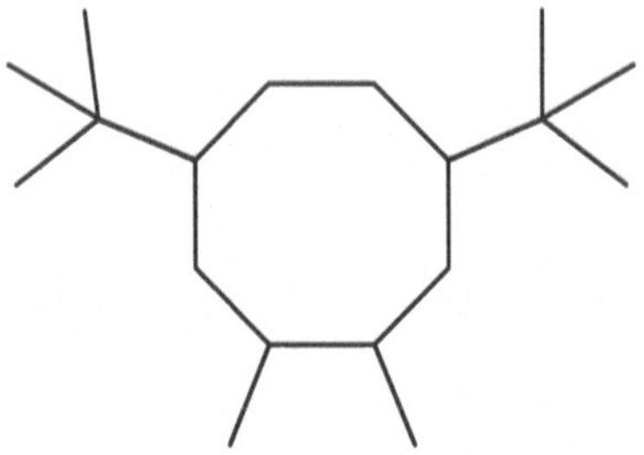

j)

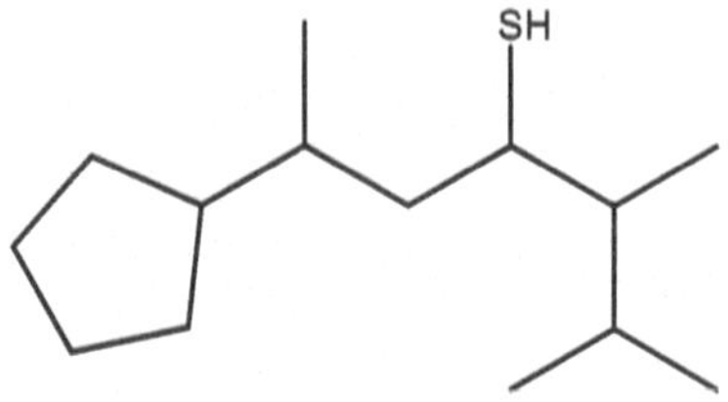
SH

k)

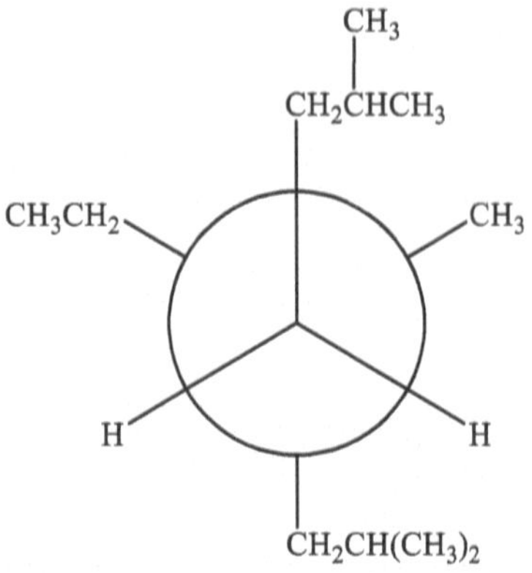
CH3
CH2CHCH3
CH3CH2
CH3
H
H
CH2CH(CH3)2

l)

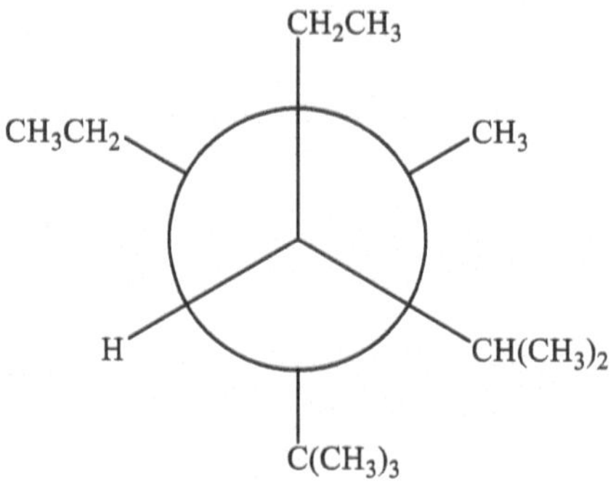
CH2CH3
CH3CH2
CH3
H
CH(CH3)2
C(CH3)3

4.2. Draw the structures corresponding to the following names

a. *trans*-4-*tert*-Butylcyclohexanol

b. (*E*)-3-Chlorohex-2-ene

c. (3*E*,5*E*)-3,5,7-Trimethylocta-1,3,5-triene

d. (1*E*,3*E*)-1-Bromo-2-methylpenta-1,3-diene

e. 4-Ethyl-5-isobutyl-6-isopropyl-2,2-dimethyldecane

f. *trans*-1-isopropyl-3-methylcyclopentane

g. 4-Ethyl-2,4,5-trimethylheptane

h. *trans*-1-*tert*-Butyl-3-methylcyclohexane

i. (*R*)-1-Phenylethanol

j. (2*E*, 4*Z*, 6*E*)-2,4,6-Octatriene

k. 2,4-Dimethyl-1,3-pentadiene

l. 4-(Dimethylamino)cyclohexan-1-one

m. 2-Ethylbutan-1-amine

n. 4-Hydroxy-3-methoxybenzaldehyde

Stereochemistry

5

What you will learn:

- Chirality
- *Cis/trans* system
- R/S system
- E/Z system
- Prochirality
- Relationships between sets of stereoisomers

5.1 INTRODUCTION

Stereochemistry is simply defined as the three-dimensional spatial arrangement of atoms in a molecule, and it is one of the most fundamental aspects of the life sciences. With our ever-increasing understanding of the natural world, we find important implications for the stereochemistry of our molecules. While understanding this importance, we often do not further explore these concepts after their initial introduction in organic chemistry but will simply mention their importance.

5.2 IMPORTANT DEFINITIONS

- **Stereochemistry:** The three-dimensional arrangement of atoms in a molecule.
- **Stereoisomers:** Isomers of a molecule that have the same molecular formula and connectivity but differ in their arrangement in space.
- **Stereocenter/chiral center:** An atom or center within a molecule, having four unique groups attached, which rotates plane polarized light.
- **Cahn–Ingold–Prelog rules:** The method by which groups are assigned priority based on atomic number.
- ***Cis/trans* system:** The systematic method by which we assign relative stereochemistry about a chiral center or double bond.
- **R/S system:** The systematic method by which we assign absolute stereochemistry about a chiral center.
- **E/Z system:** The systematic method by which we assign absolute stereochemistry about a double bond.
- **Prochiral:** A non-chiral center where chirality is introduced when one group is substituted or added.
- **Pro R/S:** The method we use to assign R/S to a prochiral stereocenter.
- **Re/Si:** The method used to describe the face of a sp^2 system (most often of a carbonyl).
- **Enantiomers:** Non-superimposable mirror image stereoisomers of a molecule, but more simply characterized as having complete inversion of all stereocenters.

- **Diastereomers:** Non-mirror image stereoisomers of a molecule, but more simply characterized as having inversion of some but not all stereocenters.
- **Meso:** A molecular with chiral centers; however, a plane of symmetry can be found within the molecule that prevents the molecular from rotating plane polarized light.

5.3 CHIRALITY

The assignment of stereocenters is a fundamental skill needed for all students of organic chemistry. This is not just a skill limited to students in an academic setting. It allows us to accurately describe a molecule so that one scientist can know exactly, without any ambiguity, what another scientist is trying to convey. Many drugs also have a single active or effective form or, worse yet, the wrong form is ineffective or toxic. A classic example involved the brand name prescription drug Contergen (generically known as Thalidomide) during the 1950s. Originally prescribed as an effective anti-nausea agent for morning sickness, many patients who took Contergen birthed babies with a range of defects. These birth defects were traced back to one of the two enantiomers of thalidomide (Figure 5-1).

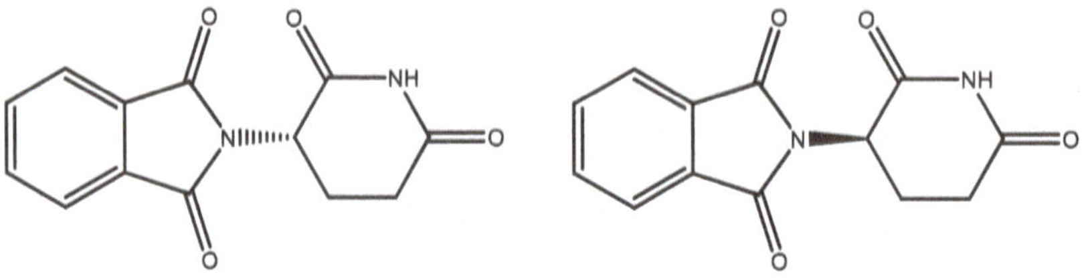

S-enantiomer R-enantiomer

Figure 5-1: Thalidomide originally sold as a mixture of R/S enantiomers was an effective medication for morning sickness. Later it was found that while the R enantiomer was effective at treating morning sickness, the S enantiomer was tetragenic. Thalidomide is still prescribed and sold today as an anti-cancer medication.

Chirality is the basis for defining stereochemistry within a molecule. A chiral molecule is defined as one that rotates plane polarized light. Since we cannot carry around plane polarized light rotation detectors, we need a more simplified way to define chirality. We can define this as an atom that has four unique groups attached to it, also known as an ABCD system (Figure 5-2). When four unique groups are attached, the molecule will rotate plane polarized light making the molecule fit the definition of chirality.

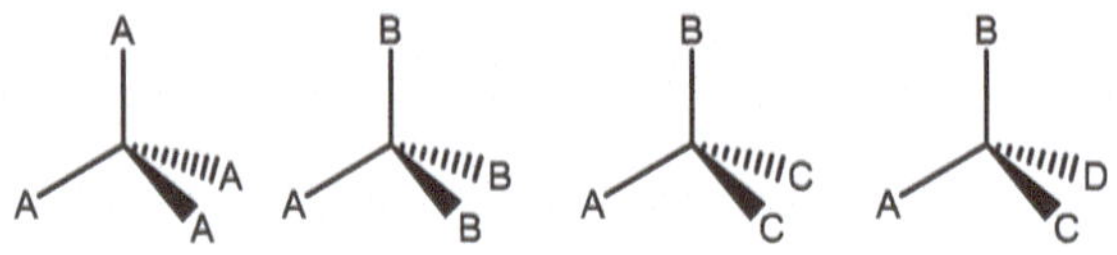

System	AAAA	ABBB	ABCC	ABCD
# of Unique Groups	1	2	3	4
Chiral	No	No	No	Yes

Figure 5-2: The ABCD system that illustrates the number of unique groups found on an atom. These range from four identical groups on the central atom (AAAA—ex methane) to four unique groups (ABCD—example 1-bromo-1-chloro-1-fluoromethane) attached to the central atom.

Practice

5.3.1. Indicate with an asterisk the location of stereocenters

a)

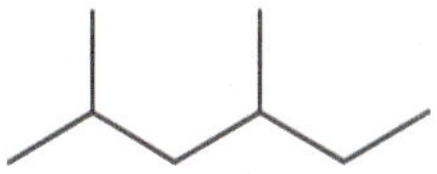

b)

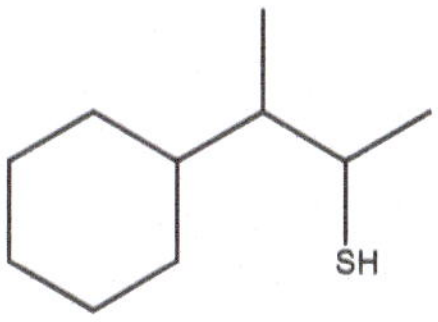

c)

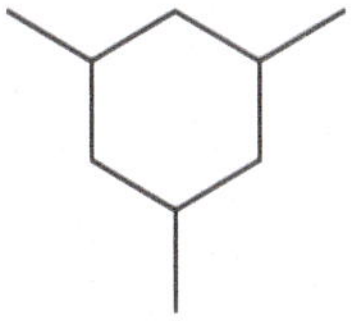

d)

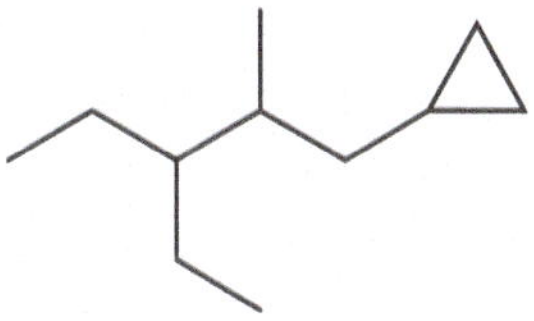

e)

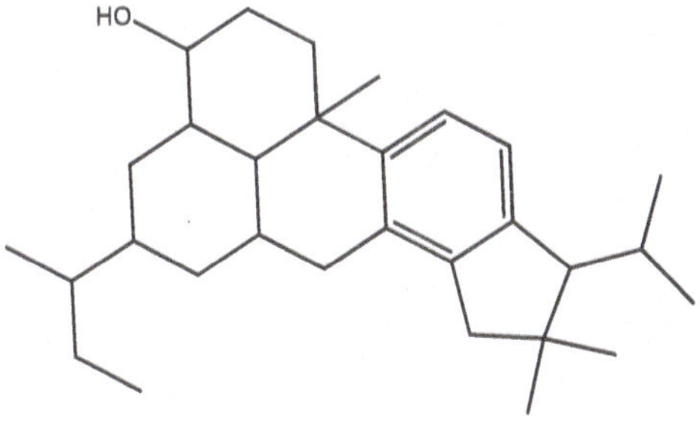

f)

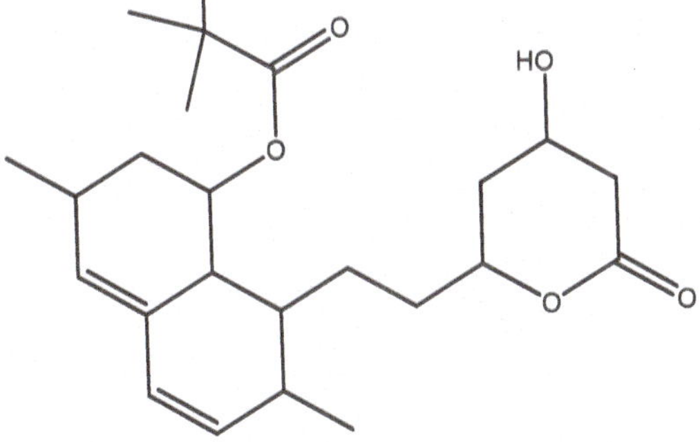

g)

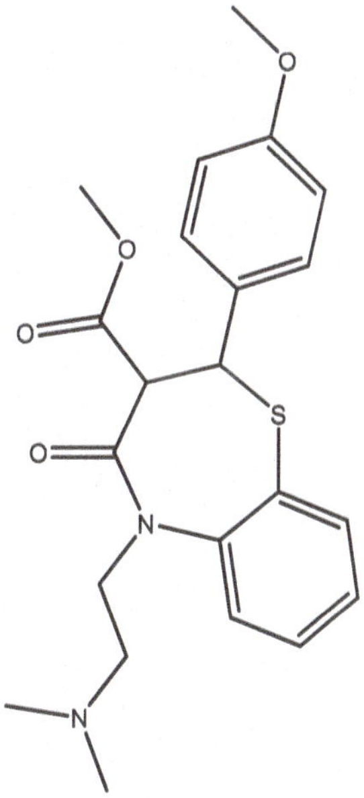

5.3.2. How many stereocenters do each of these molecules have?

a)

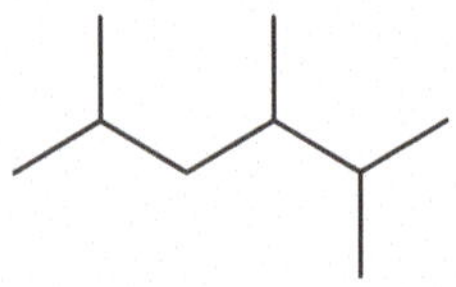

b)

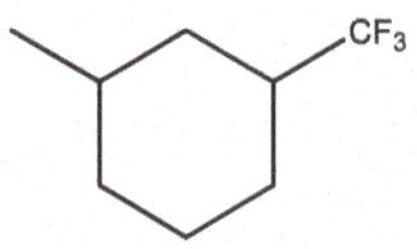

c)

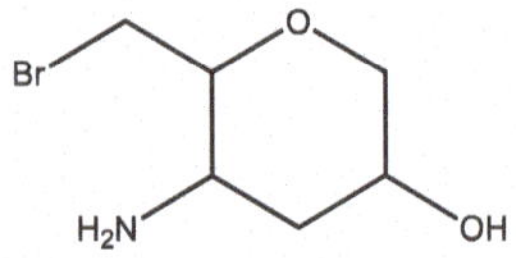

d)

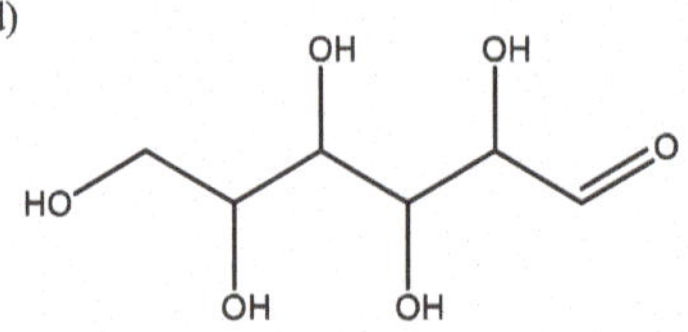

e)

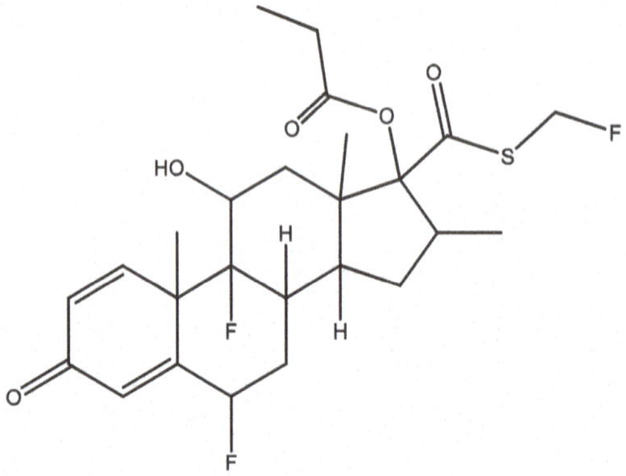

f)

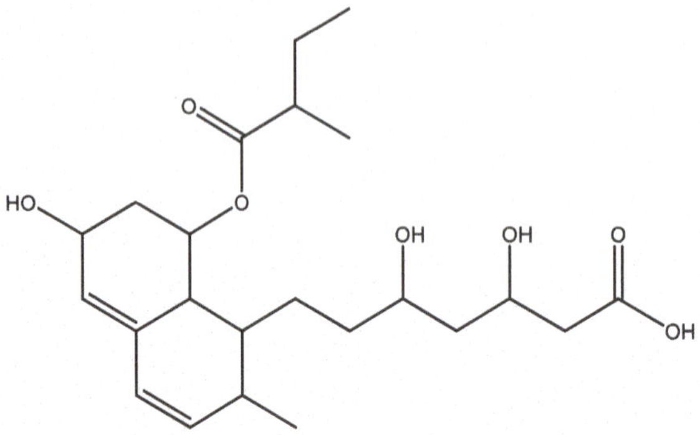

5.3.3. Indicate whether each of these molecules are chiral or achiral

a)

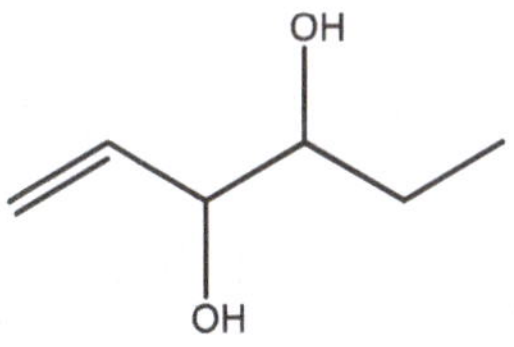

b)

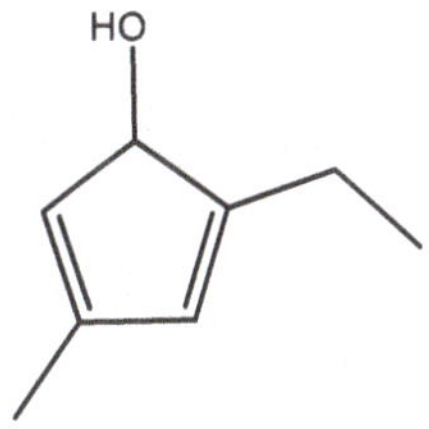

c)

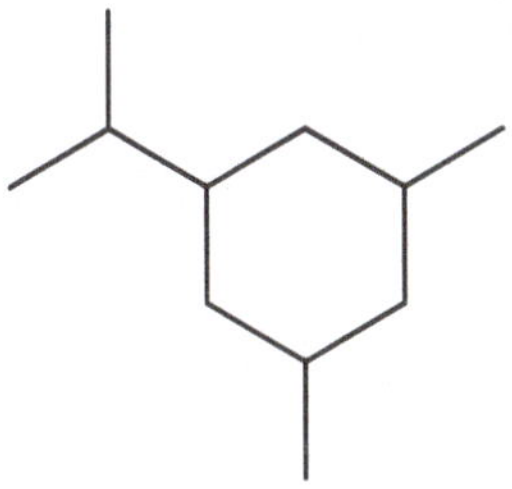

d)

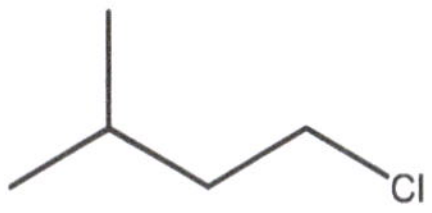

e)

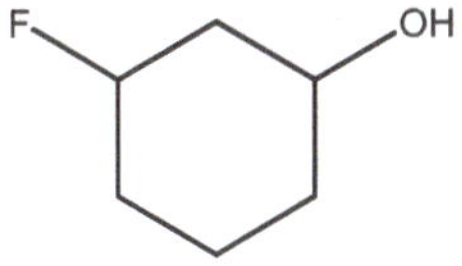

f)

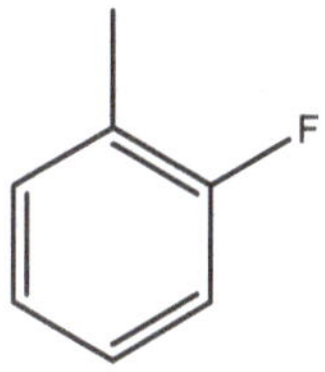

g)

h)

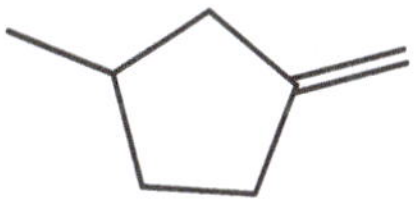

5.4 CIS/TRANS SYSTEM

The cis/trans system is the method we use to describe relative relationships between two groups; this is not an absolute descriptor like some of our later methods. This convention can be used in describing sp^3 (acyclic and cyclic systems) and sp^2 systems (double bonds) allowing the relation of any two groups in three-dimensional space on the same molecule (Figure 5-3). For the sp^3 systems we determine whether two stereocenters are in the same orientation as they relate to how the molecule is drawn. It is important to note that we have no information about a specific stereocenter itself, just its illustrated relationship to another stereocenter. In the case of sp^2 systems, we can only use the cis/trans convention to compare the orientation of groups on disubstituted, non-terminal double bonds (on the same side or opposite side of the double bond). It is important to realize that these are relative terms and are not used in naming the compound for formal purpose (IUPAC naming rules), except when no assignment of absolute stereo chemistry is available. We also do not use cis and trans to describe the relationship of a double bond within rings smaller than eight members as they are unable to adopt the trans configuration due to bond angle restraints.

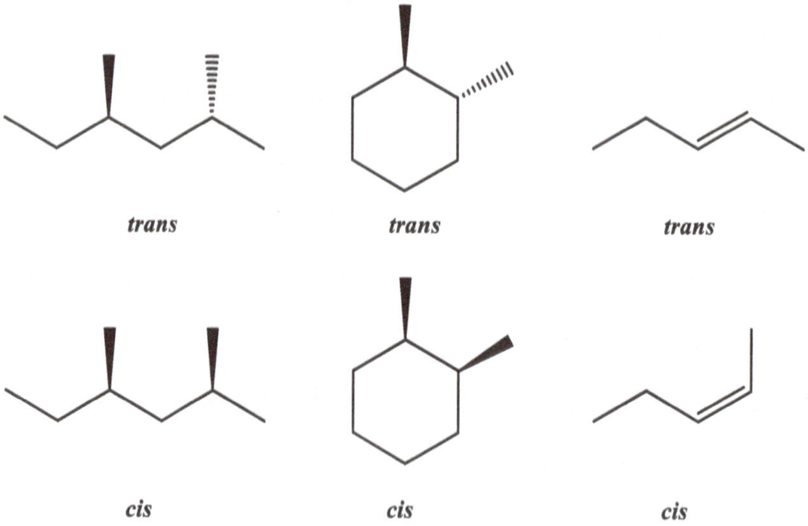

Figure 5-3: ***Selection of possible relative stereochemical situations.***

Practice

5.4.1. Use the *cis/trans* system to indicate relative stereochemistry for each of these molecules!

a)

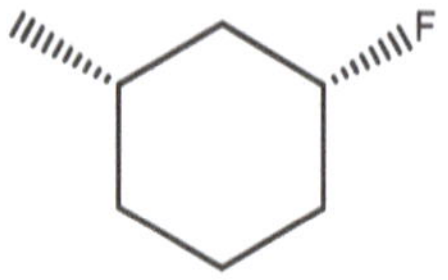

b)

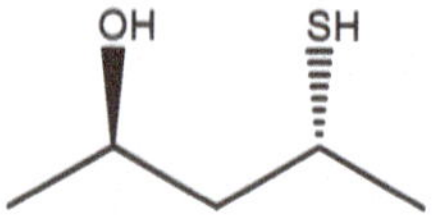

c)

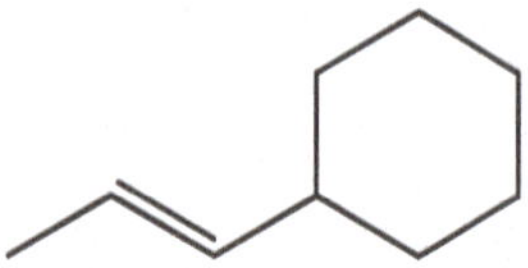

d)

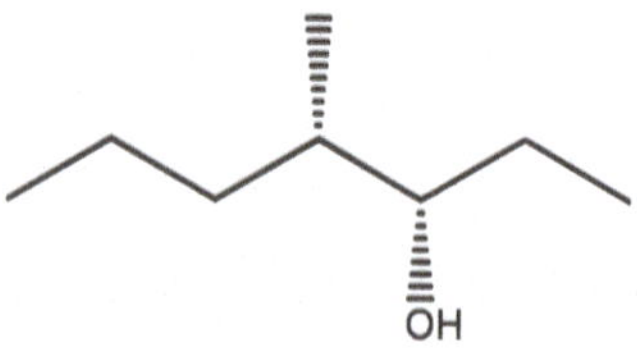

e)

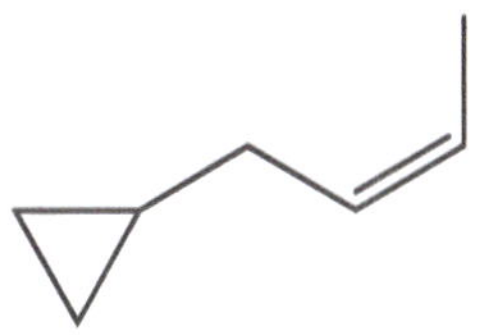

f)

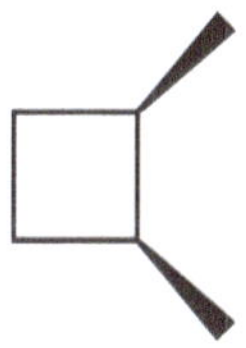

g)

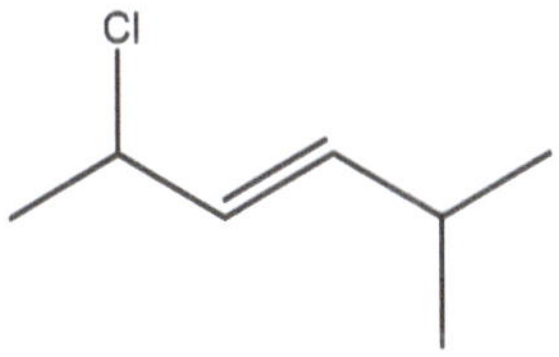

h)

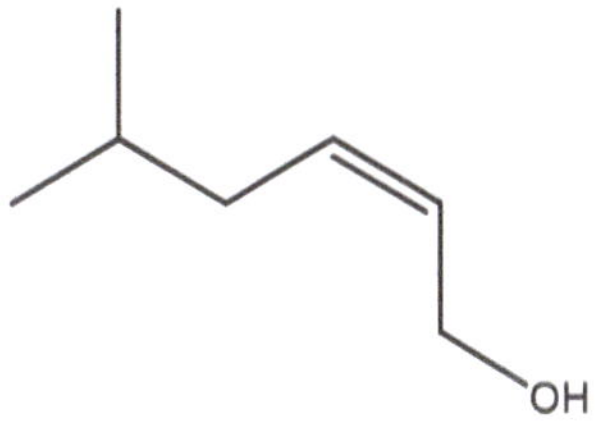

5.5 R/S SYSTEM

R and S are the systematic descriptors we use to communicate information about a specific stereocenter in a molecule. When assigning R/S we look at the molecule and assign priority to each unique group (the higher the atomic number the higher the priority) and investigate. With priority 4 facing away from us (Figure 5-4—Case 1), if groups 1 through 3 are oriented in a clockwise pattern (example A), the stereocenter is in the R configuration; If they are in a counterclockwise pattern (example B), the stereocenter is in the S configuration. If priority 4 is facing toward us (Figure 5-4—Case 2), we invert the stereocenters to account for how we have drawn it. Looking at Case 2, example C, we then see that groups 1 through 3 are oriented in a clockwise pattern, indicating an R configuration, which is then inverted to the S configuration. However, the groups in example D are oriented in a counterclockwise pattern yielding the S configuration, which is then inverted to the R configuration.

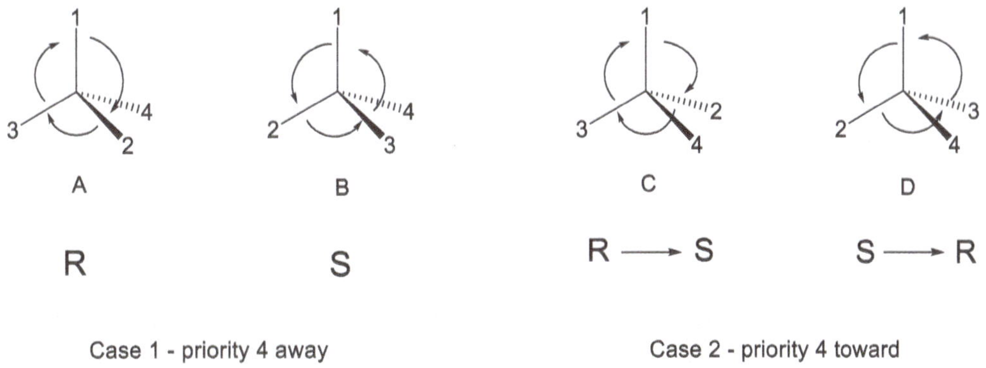

Figure 5-4: The R/S system, showing the dependence on how the molecule is drawn.

Practice

5.5.1. Use the R/S system to indicate absolute stereochemistry in the following molecules!

a)

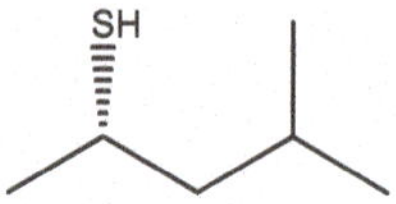

b)

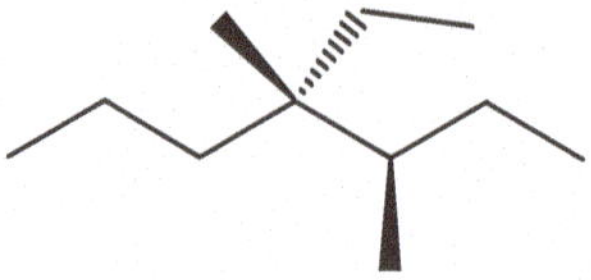

c)

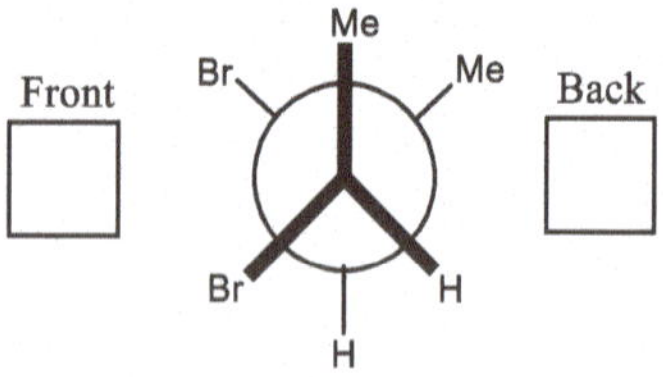

d)

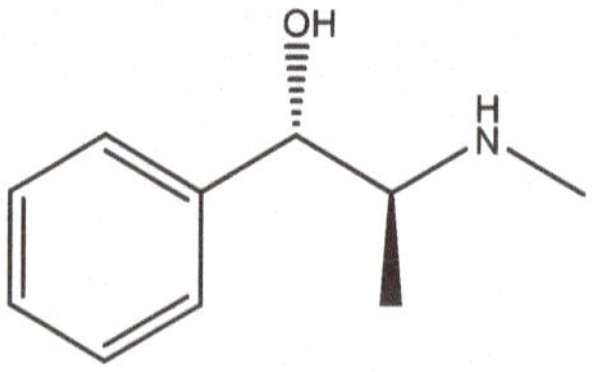

e)

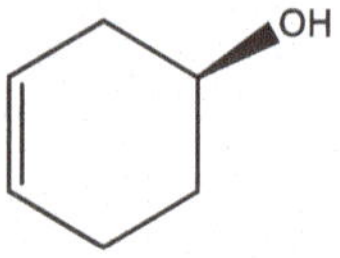

f)

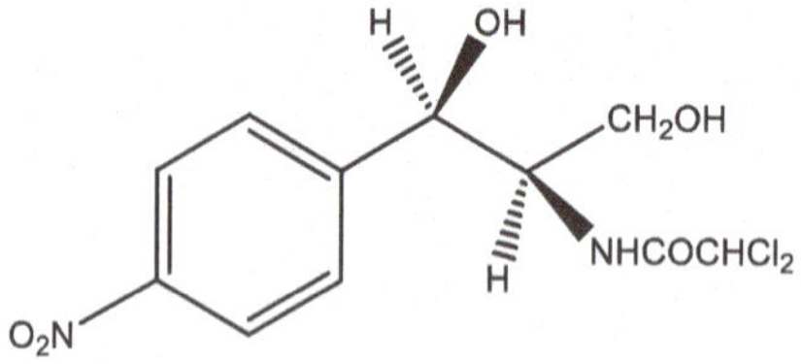

g)

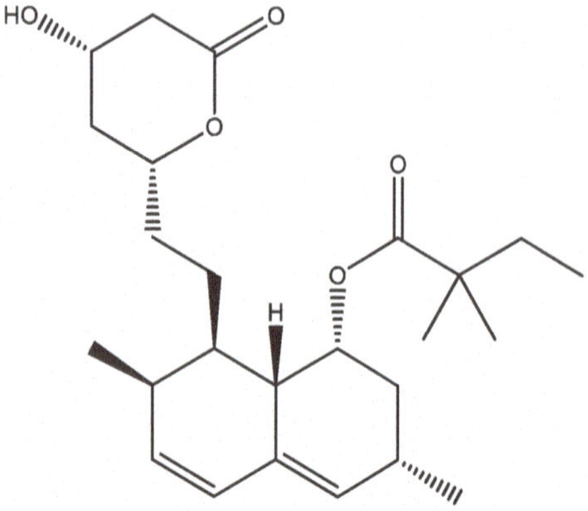

h)

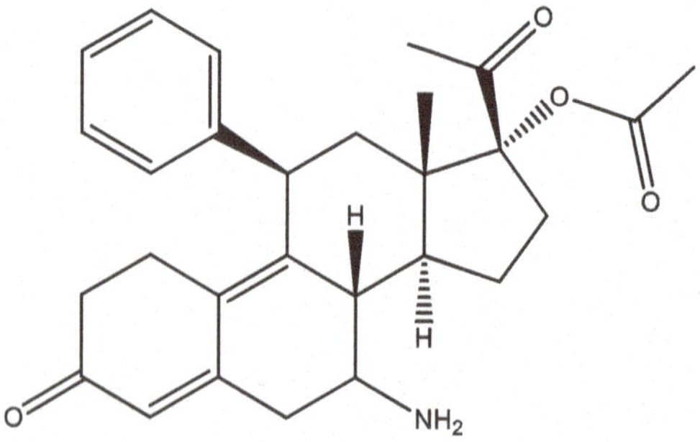

5.5.2. Draw the line angle structure of the names below!

a. (*R*)–2,3-Dimethyloctane

b. (4R,5S)-4-ethyl-2,4,5-trimethylheptane

c. (*S*)-3-chloro-2-methylpentane

d. (1*R*,3*R*)-1-isopropyl-3-methylcyclohexane

e. (*S*)-3-ethyl-2,5-dimethylhexane

f. (2*R*,4*R*)-2-chloro-4-fluoro-4-methylhexane

5.6 E/Z SYSTEM

E and Z are the systematic descriptors we use to communicate information about the stereochemistry of a sp^2–sp^2 system, specifically the double bond. The assignment of E/Z is similar to the assignment of R/S. First, it is necessary to make a bilateral section of the double bond (Figure 5-5). Each side of the double bond (labeled a and b) is treated independently. Next assign the lowest and highest priority for each group on both sides. The Z isomer will have both of the groups on the same face (this is a cis relationship) while the E isomer will have the groups on opposite faces of the double bond (this is a trans relationship).

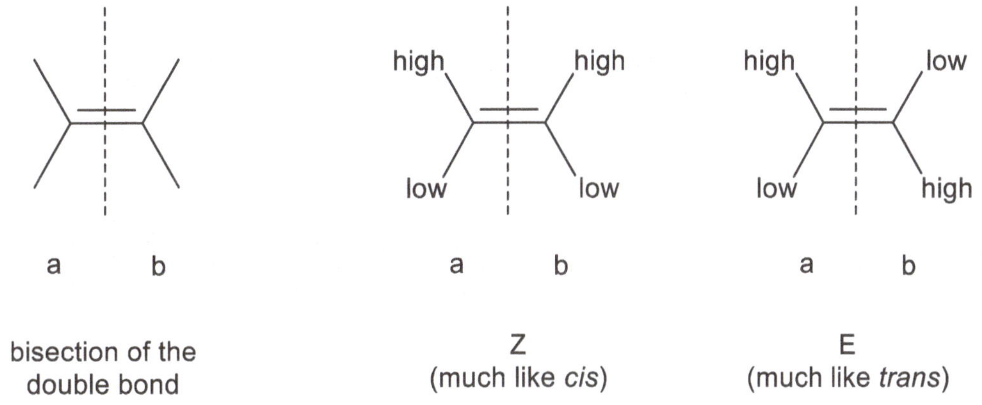

Figure 5-5: The E/Z system used to define absolute stereochemistry for sp^2 systems.

Practice

5.6.1. Use the E/Z system to indicate stereochemistry in the following molecules!

a)

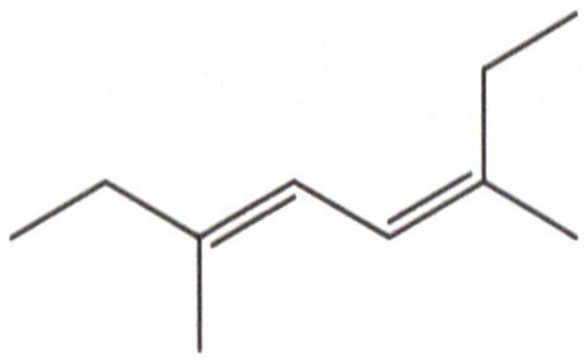

b)

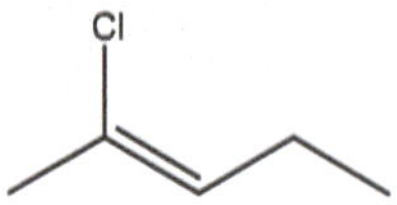

c)

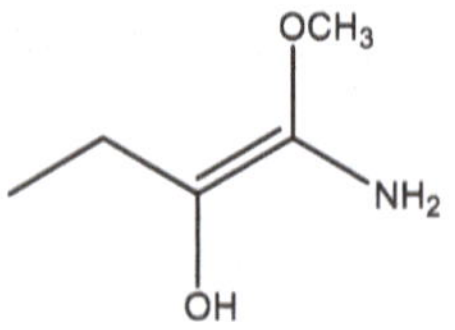

d)

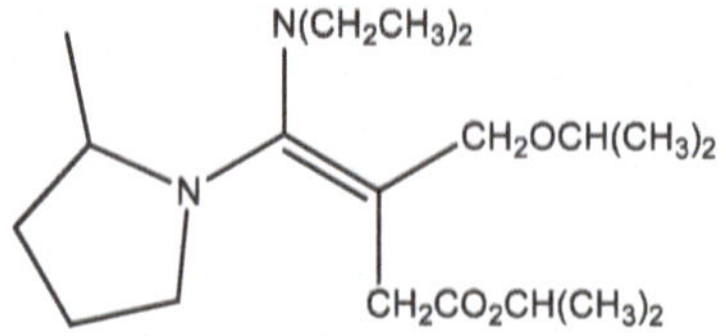

e)

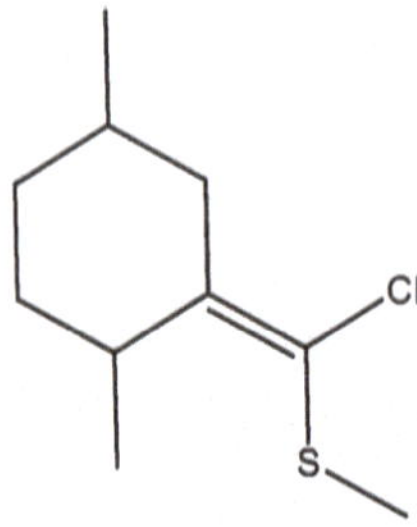

f)

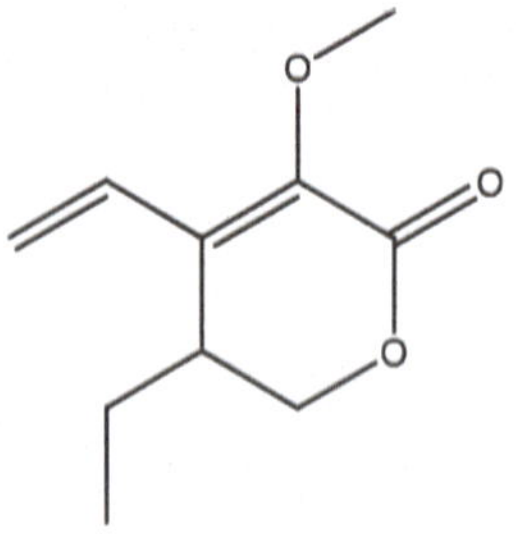

g)

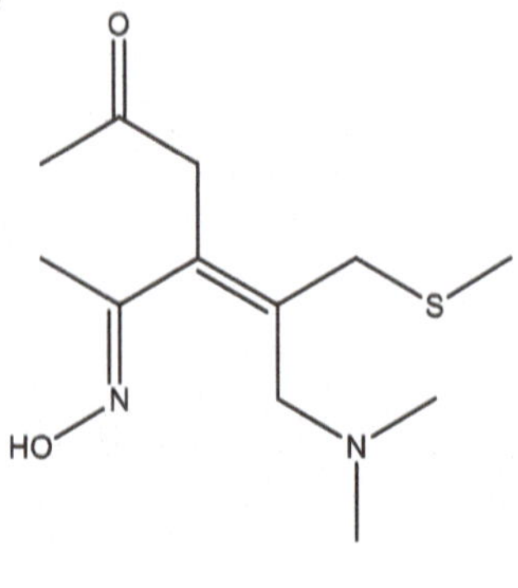

h)

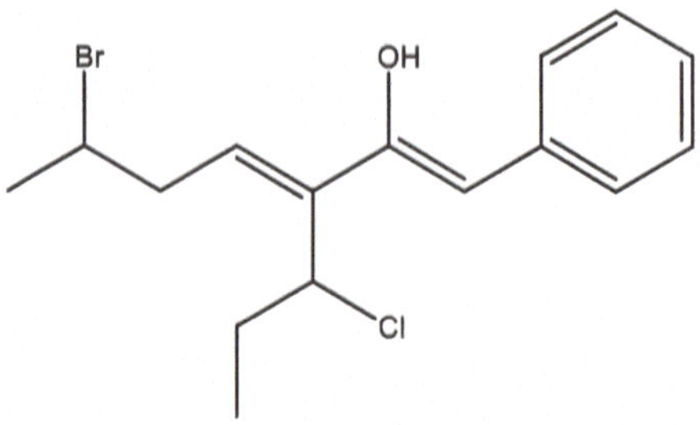

5.7 PROCHIRALITY

Even if a molecule is not chiral, this doesn't mean that it will not exhibit important stereochemical properties, we call this phenomenon prochirality. Prochirality is a trait exhibited by a molecule that upon substitution (sp^3) or addition (sp^2) of a group, the molecule becomes chiral. The most common example is found with the sp^3 ABCC-type system (Figure 5-2), usually when the two identical groups (CC) are hydrogens. Upon replacement of a hydrogen with deuterium, the central atom then becomes an ABCD system and therefore chiral. Because of this, the hydrogens get special designations Pro R or Pro S and are referred to as enantiotopic or diasterotopic hydrogens (depending on whether they create enantiomeric or diastereomeric pairs upon substitution). If replacement of the specific hydrogen causes the atom to become R, it is designated as the Pro R hydrogen. Similarly, if the replacement of the specific hydrogen causes the atom to become S it is designated as the Pro S hydrogen (Figure 5-6). The same situation exists among sp^2 systems, whereby the addition of a group induces chirality depending on which face the group adds. A typical example involves a carbonyl (or one of its analogs), which is given the designation Re and Si. To assign Re and Si we must determine the priority of the groups attached to the carbon on the carbonyl (priorities 1–3). A Re face designation is given when a clockwise orientation is present and a Si face designation is given when the orientation is counter-clockwise.

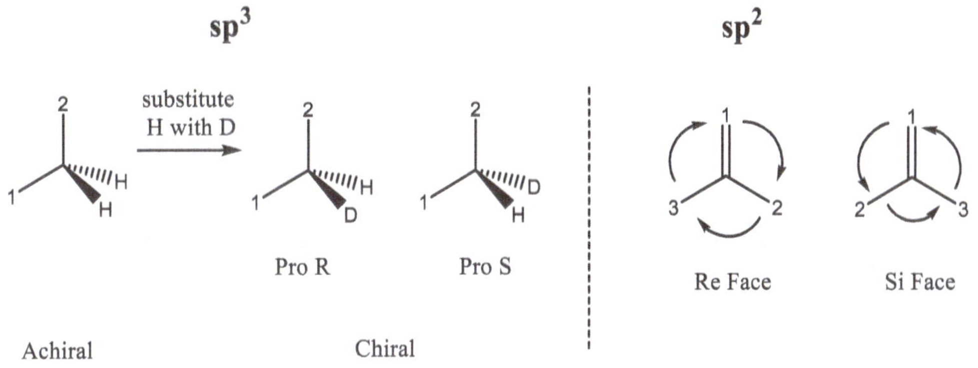

Figure 5-6: Prochirality demonstrated in both sp^3 and sp^2 systems.

Practice

5.7.1. Assign ProR or ProS to the prochiral hydrogens in the following molecule!

a)

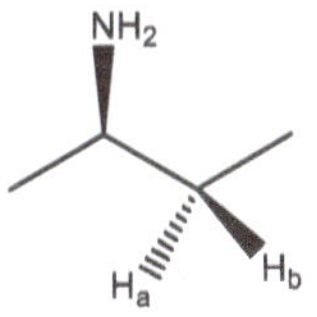

b)

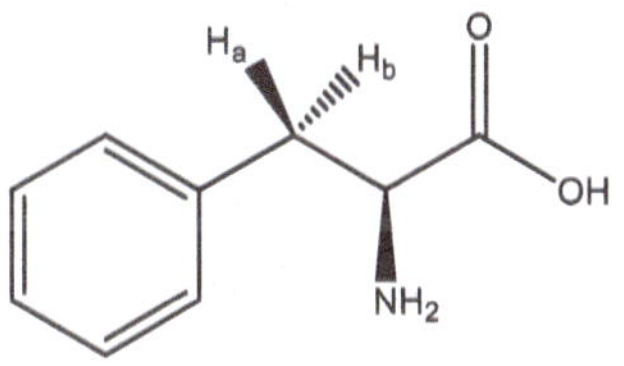

c)

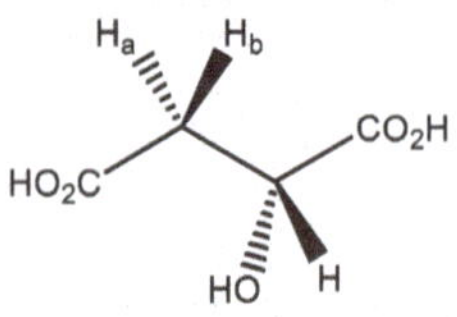

d)

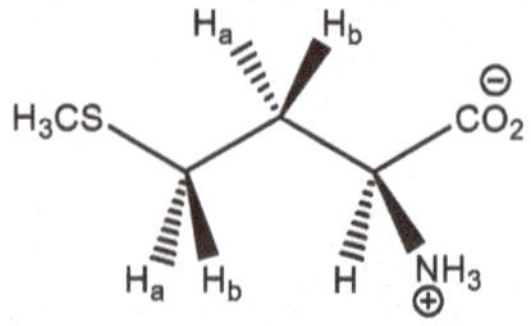

e)

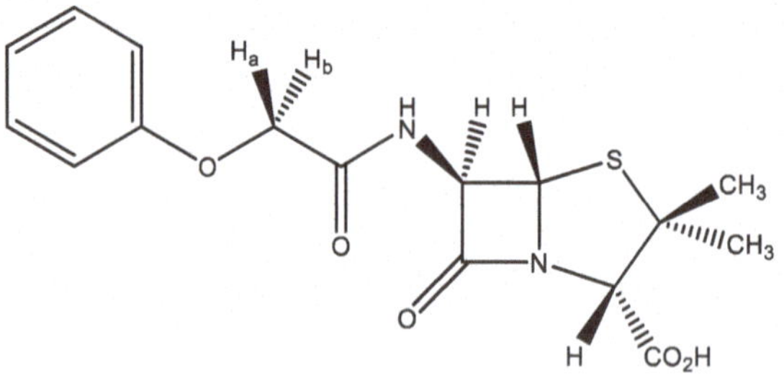

5.7.2. Indicate Re and Si face in the following molecules!

a)

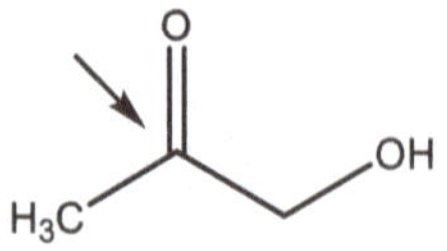

b)

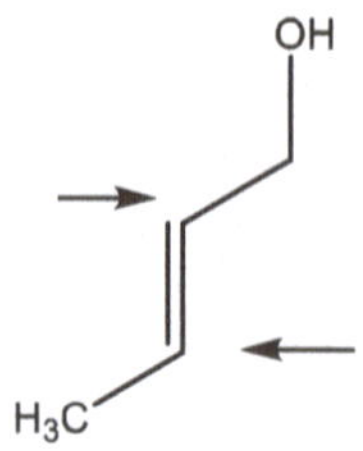

c)

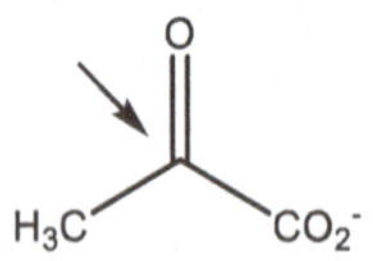

d)

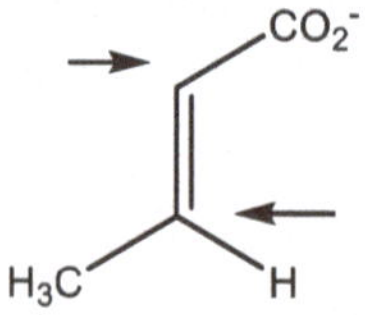

e)

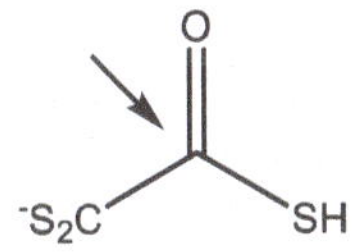

f)

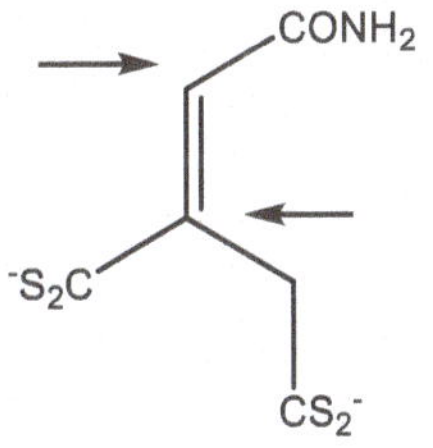

5.8 RELATIONSHIPS BETWEEN SETS OF STEREOISOMERS

While the previous examples have discussed the assignment of stereocenters, we will now investigate the relationship among sets of stereoisomers. First we will determine the possible number of stereoisomers in a molecule, using the formula $2n$, where n = the number of chiral centers (possible stereocenters). While this number gives us the total possible number of stereocenters (think maximum number of stereocenters), it does not guarantee that this number of isomers will be observed. For example, cholesterol has eight stereocenters so the possible number of isomers (28) is 256. However, it is interesting to see the specificity of the human body and that it will only use and recognize one variant of the molecule (Figure 5-7).

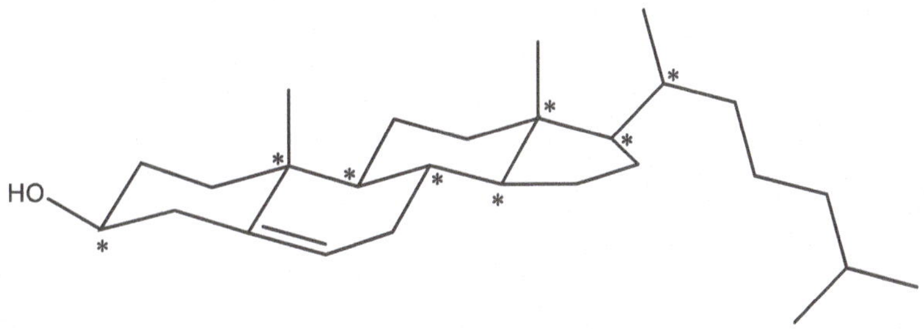

Figure 5-7: ***Cholesterol drawn from a side-on view indicating the possible stereocenters.***

Enantiomers and diastereomers are some of the most confusing of the relationships described in stereochemistry, but they need not be confusing. The confusion stems from taking the "textbook definitions" associated with these terms and trying to imagine the molecules in three-dimensional space in one's head. A simple method for doing this is to assign stereocenters and compare the results. When looking at three-aminobutan-2-ol (Figure 5-8), if all the stereocenters are inverted ((2*R*,3*S*) compared with (2*S*,3*R*)), then the relationship between the two molecules is enantiomers. Looking at the same example, if some of the stereocenters are inverted, ((2*R*,3*R*) compares with (2*S*,3*R*)), then the relationship between the two molecules is diastereomers. A couple of things to note: (1) the $2n$ rule gave us four possible stereoisomers and all four possible orientations in nature and (2) being able to draw the possible stereoisomers is very important in order to check for planes of symmetry.

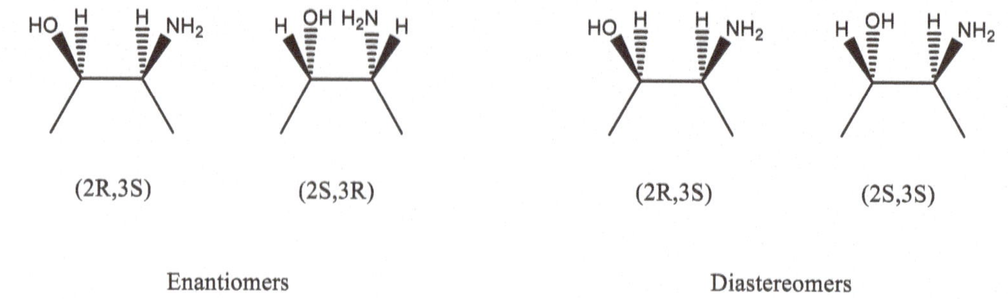

Figure 5-8: ***Illustration of the relationships between enantiomers and diastereomers for 3-aminobutan-2-ol.***

If the relationship between diastereomers and enantiomers was not confusing enough, we also need to take into consideration mesomeric isomers (meso compounds). This phenomenon sometimes occurs among enantiomeric pairs. In mesomeric isomers we find a molecule with chiral centers that are bisected with an internal mirror plane of symmetry. These compounds therefore do not rotate plane polarized light and are achiral molecules even though they have chiral centers. These two compounds are identical due to their superimposable nature. For example, tartaric acid, can form meso compounds and therefore does not for maximum number of possible stereoisomers (Figure 5-9). Examination of the meso pair formed by tartaric acid reveals a bisected internal line of symmetry with (2*R*,3*S*) and the enantiomer (2*S*,3*R*). Therefore, these two compounds are identical (a 180° rotation about the *x*-axis), and the total number of stereoisomers is reduced to three out of the possible 4 ($2n$ rule).

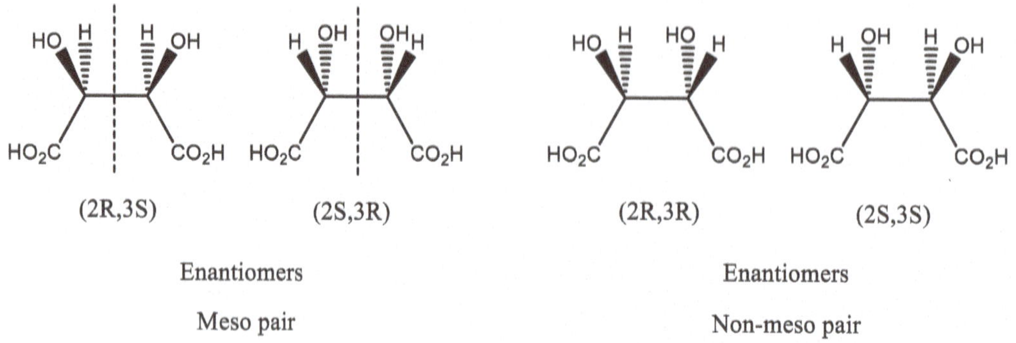

Figure 5-9: ***Stereoisomers of tartaric acid showing why there are the three possible isomers instead of four, due to the superimposable meso pair.***

Practice

5.8.1. Determine the relationship between the following pairs of molecules as identical, enantiomer, diastereomers, meso, constitutional isomer!

a)

b)

c)

d)

e)

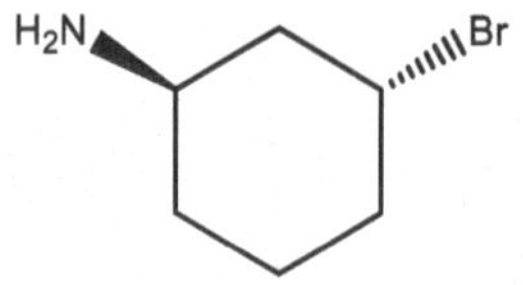

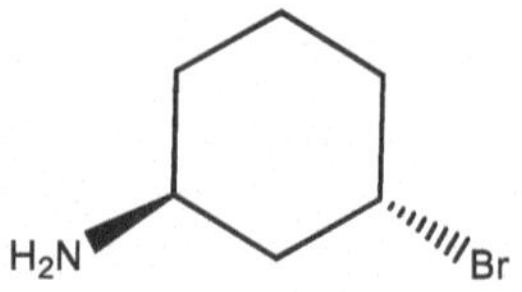

f)

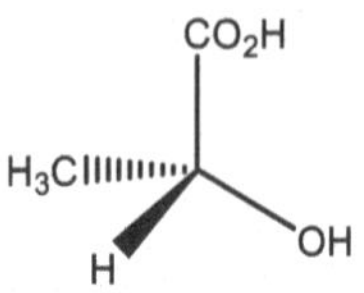

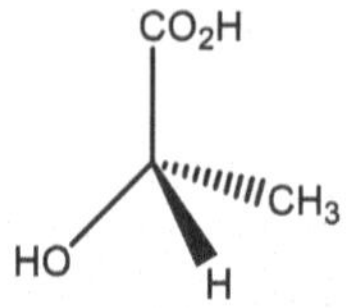

g)

Name: ______________________________ Date: ____________

INTEGRATED PROBLEMS

5.1. Provide IUPAC names for the following compounds and indicate stereochemistry where appropriate.

a)

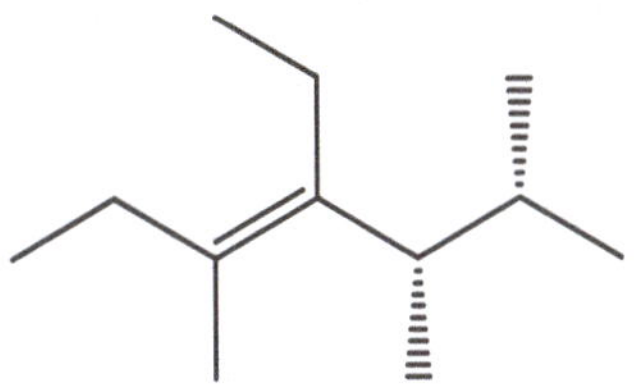

b)

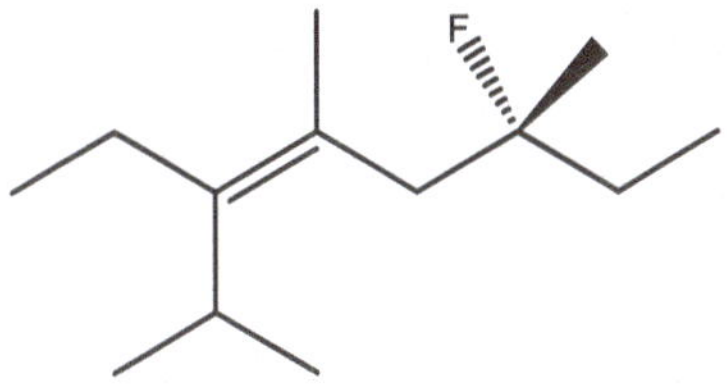

c)

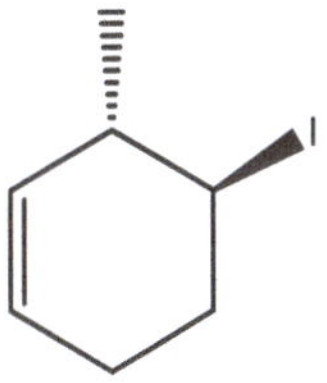

d)

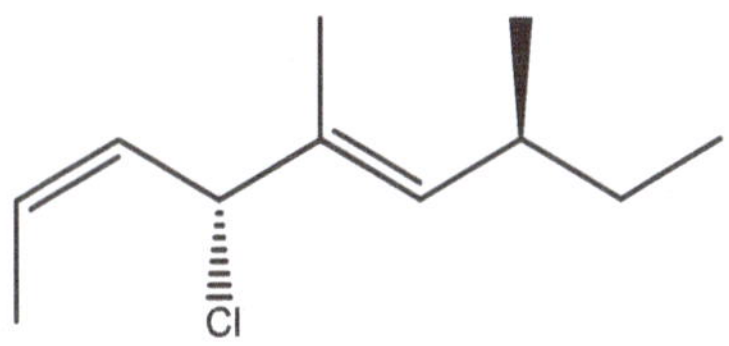

e)

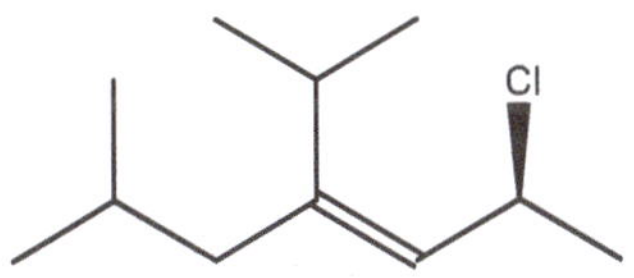

f)

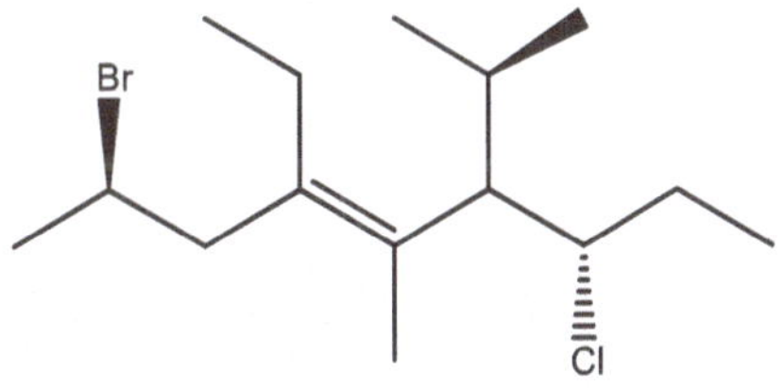

5.2. Draw the line angle structure of the names below.

a. (3*E*,5*E*)-3,5,7-trimethyl-1,3,5-octatriene

b. (3E,5Z)-3,5,9-trimethyl-3,5-decadiene

c. (4*R*,5*R*)-4-chloro-5-methylcyclohex-1-ene

d. (5*R*,6*R*,*E*)-6-chloro-5-methylhept-2-ene

e. (2*E*,4*E*,6*R*,7*R*)-7-chloro-4-ethyl-6-isopropylocta-2,4-diene

f. (2*Z*,4*E*,6*R*,7*S*)-4-ethyl-6-isopropyl-3,7-dimethyldeca-2,4-diene

5.3. Indicate absolute stereochemistry where appropriate.

a)

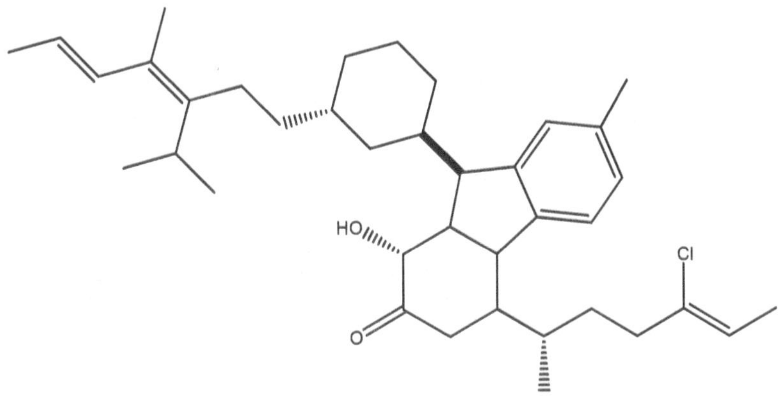

b)

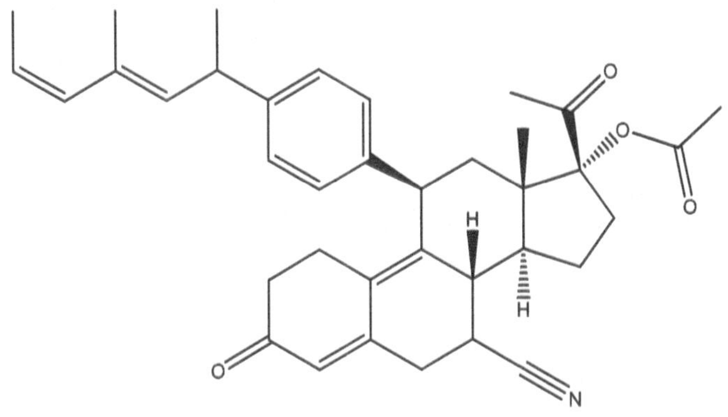

c)

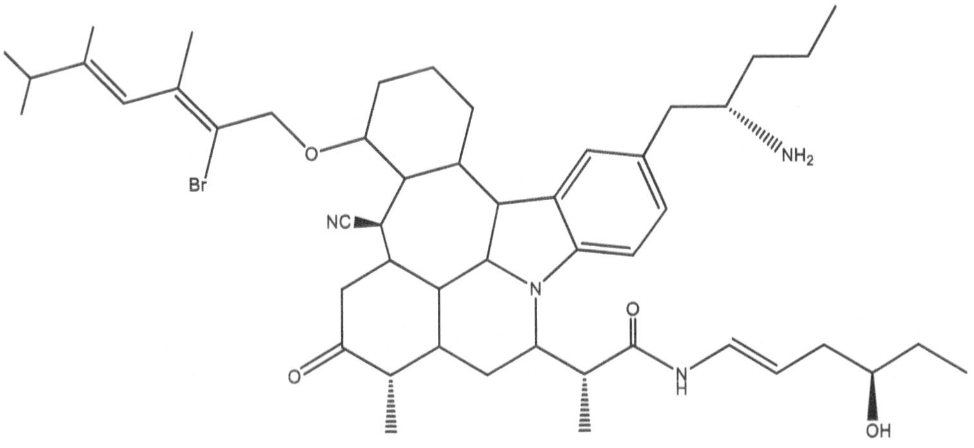

5.4. Indicate the relationship of the molecules listed to compound 1 as enantiomer, diastereomers, identical, and constitutional isomer.

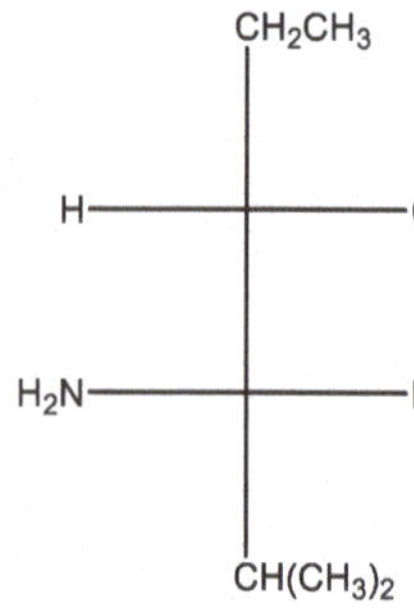

Compound 1

a)

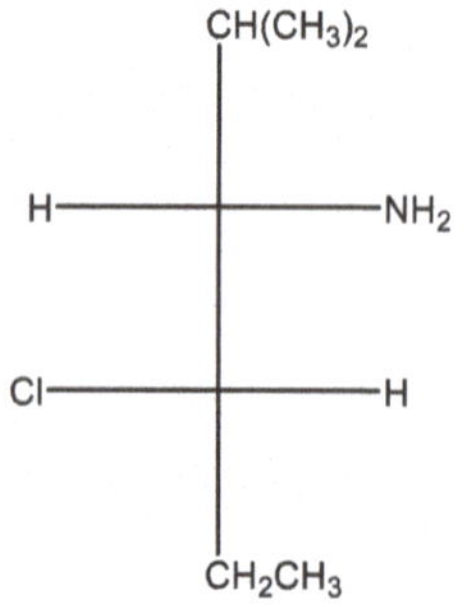

b)

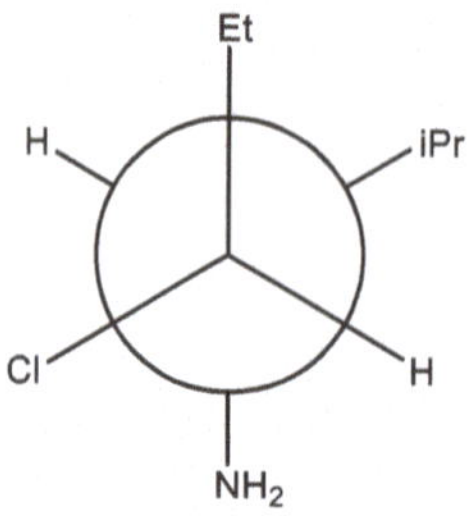

c)

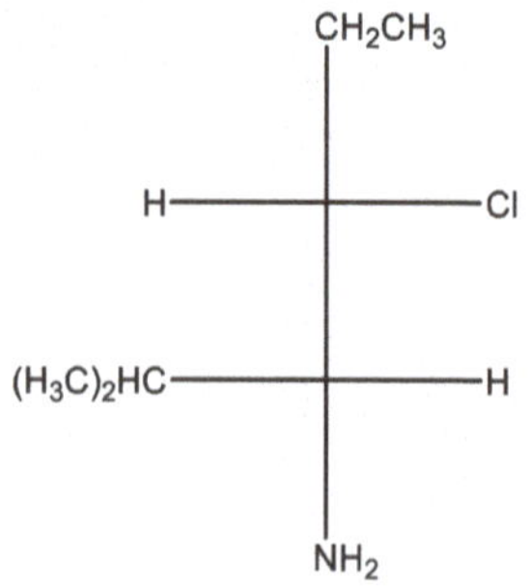

d)

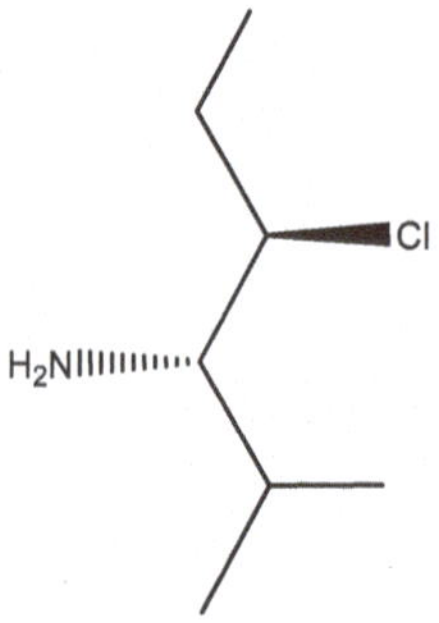

e)

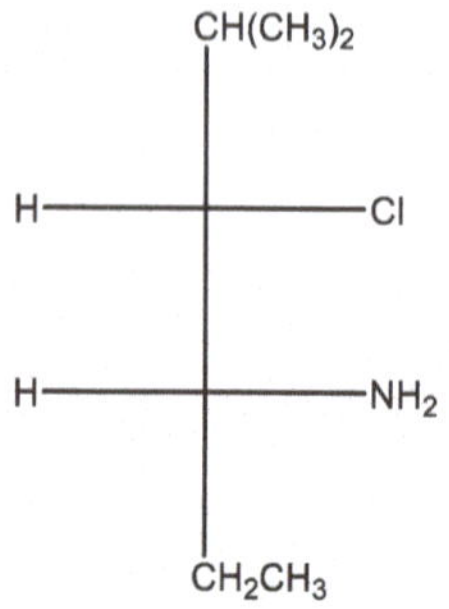

5.5. In the boxes provided below, write the relationship (**enantiomer, diastereomer, identical, constitutional isomer**) of the pair of molecules.

Appendix

SOLUTION TO 1.3.1

a) 

b) 

c) 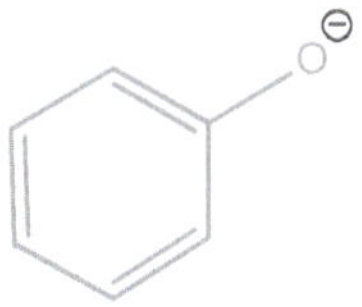

d) 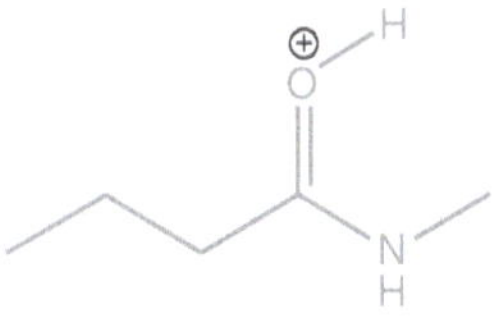

e) 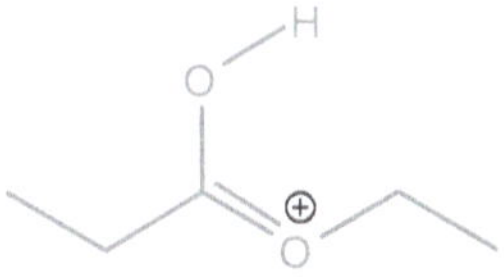

f) 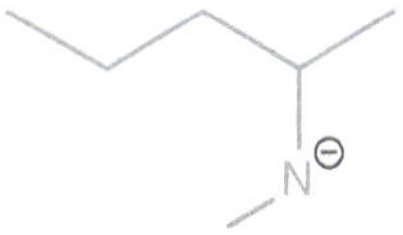

SOLUTION TO 1.4.1

a) 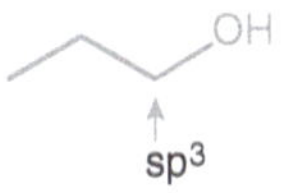

b) 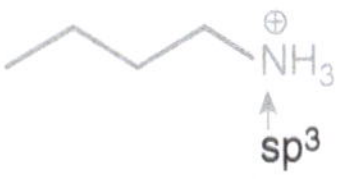

c) 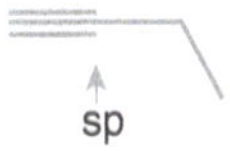

d) 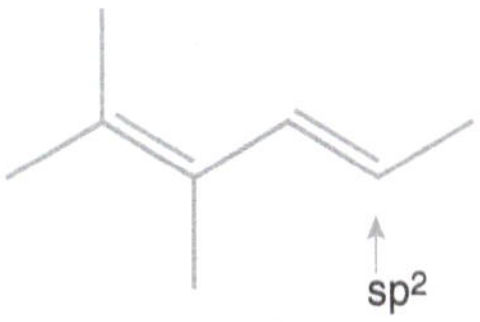

e)

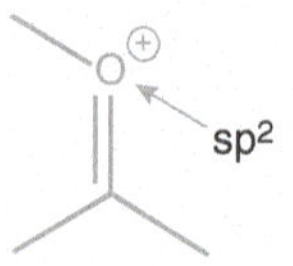

f)

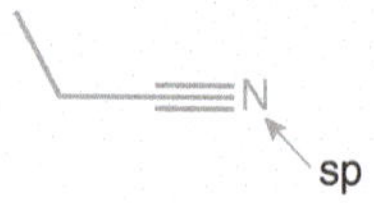

g)

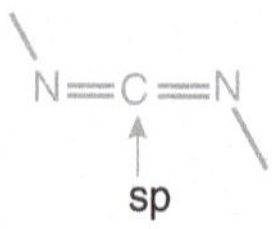

h)

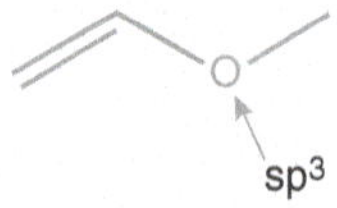

i)

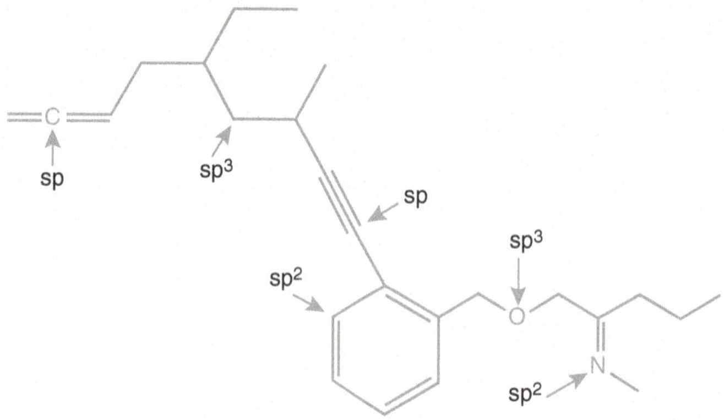

SOLUTION TO 1.6.1

a)

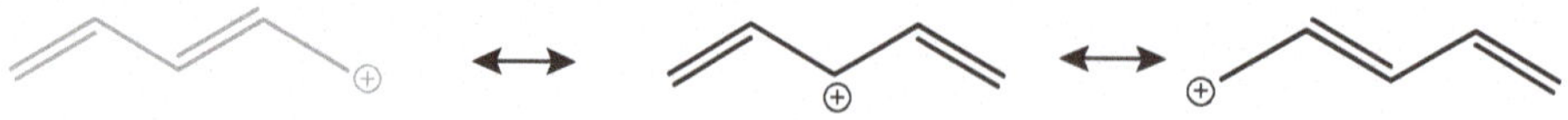

b)

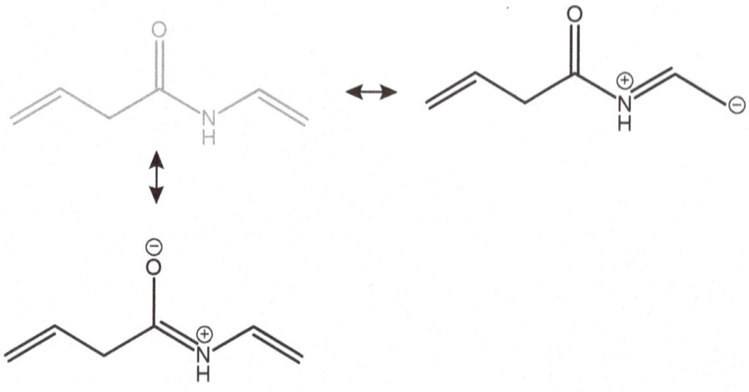

c)

d)

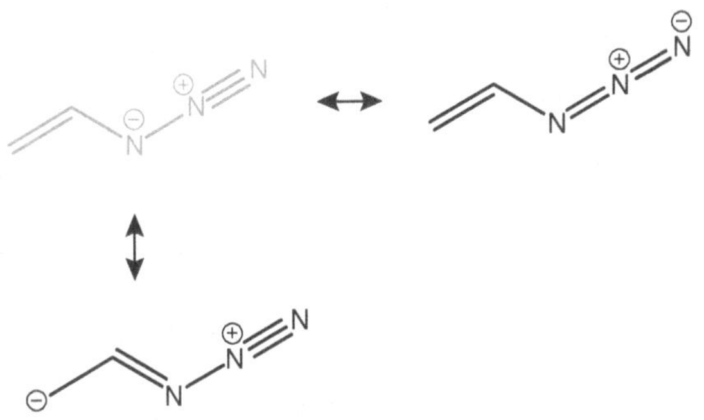

e)

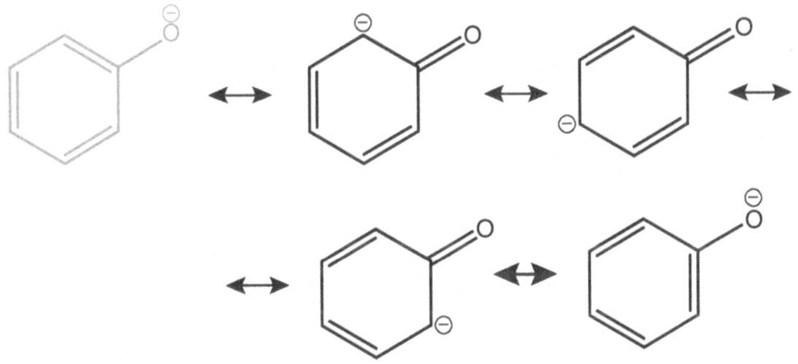

f)

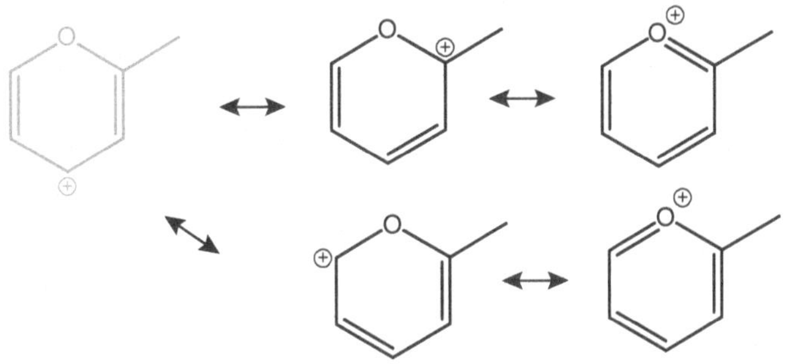

g)

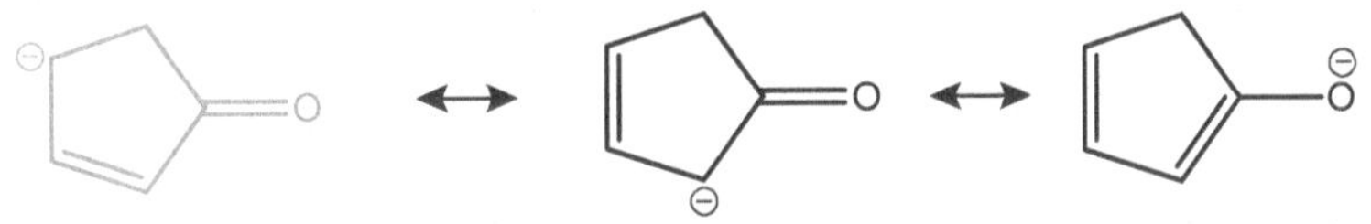

h)

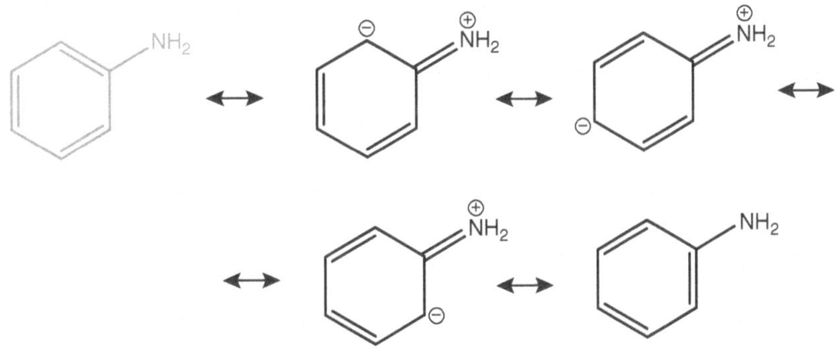

SOLUTION TO 1.7.1

a)

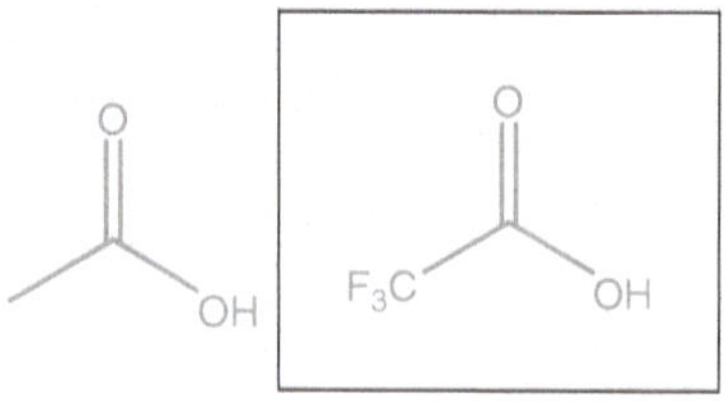

b) H-Cl H-F

c) NH_3 NH_4Cl

d) H_2O $HO^{\ominus}$

SOLUTION TO 1.7.2

a)

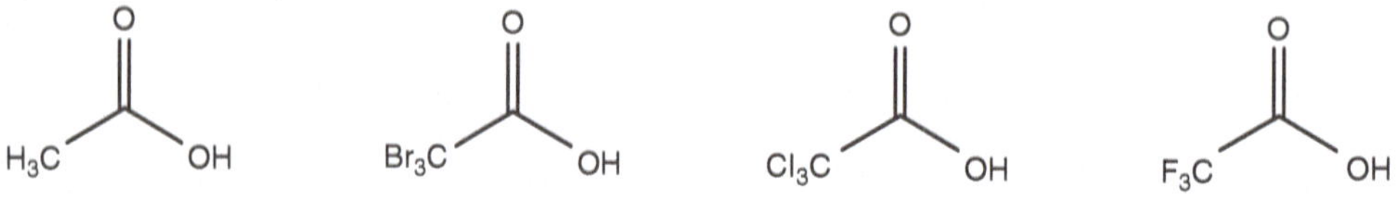

b)

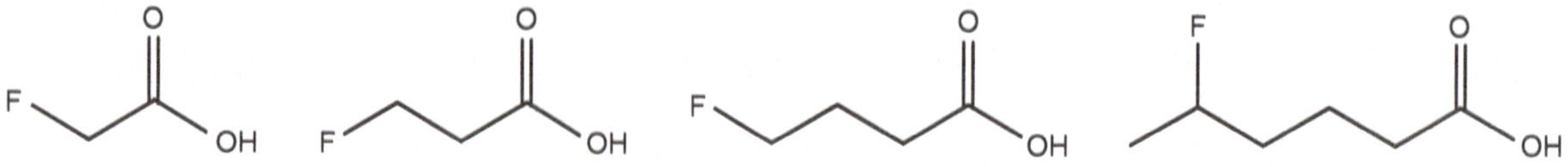

c)

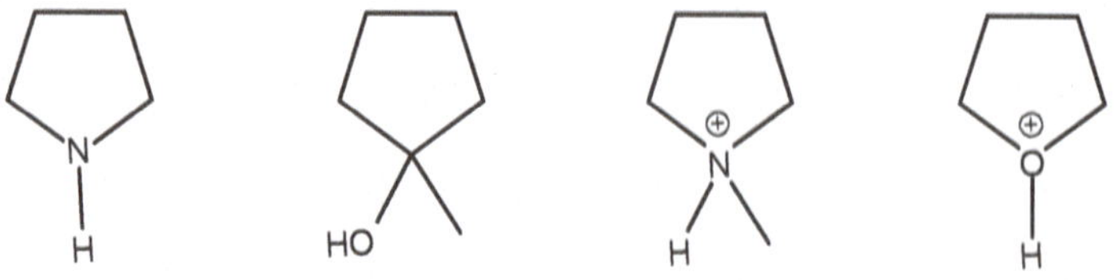

d)

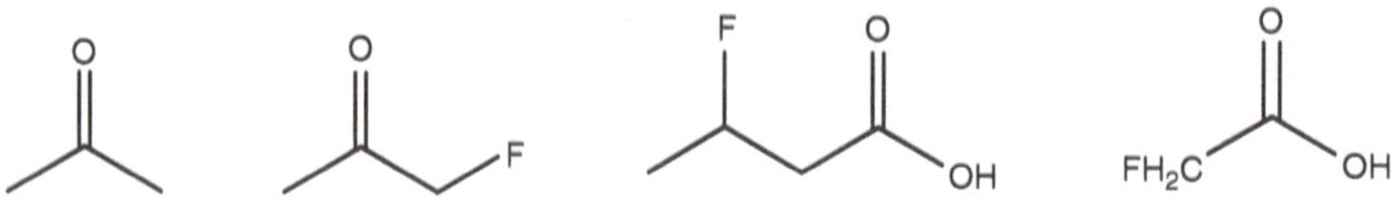

e)

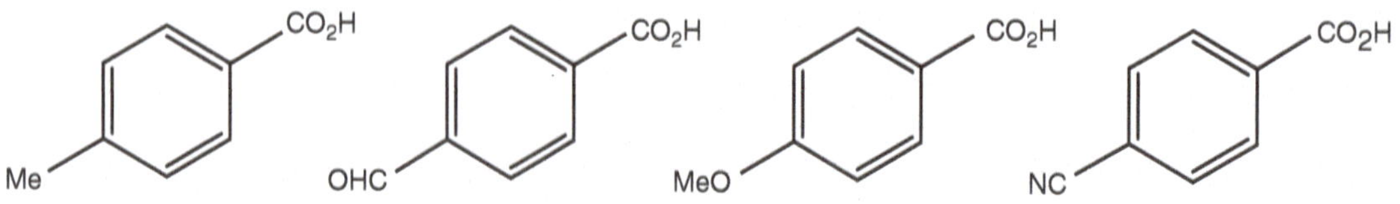

SOLUTION TO 1.7.3

a)

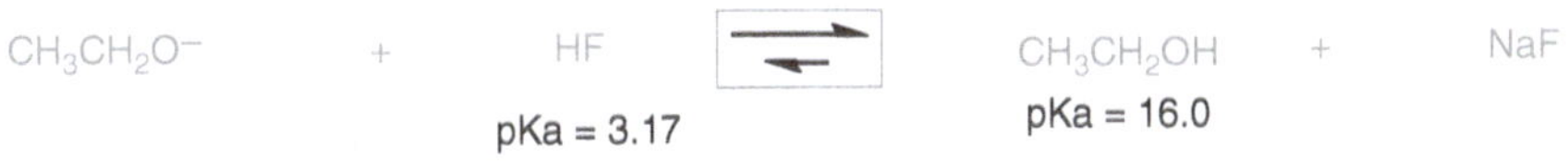

b)

c)

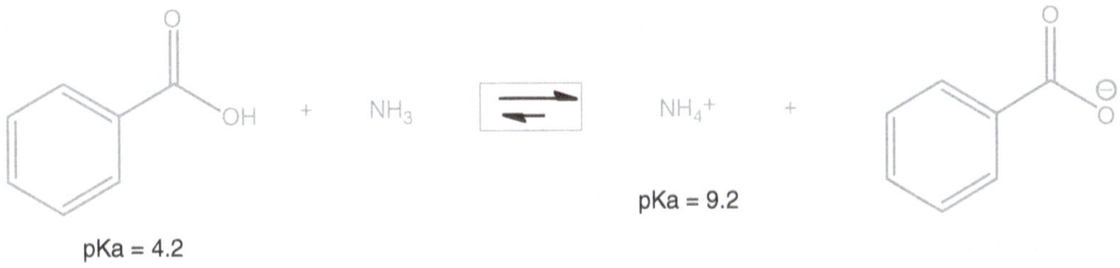

d)

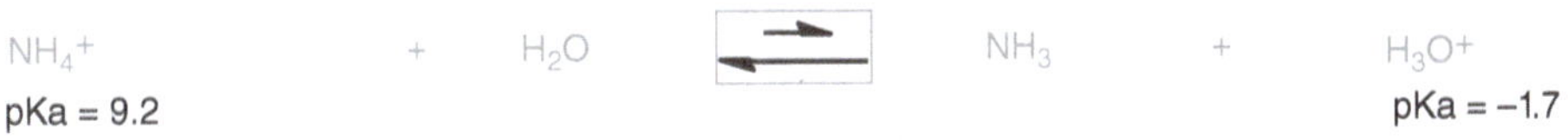

e)

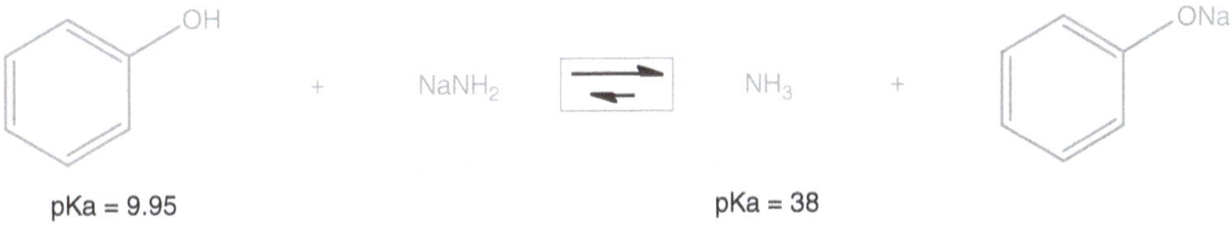

SOLUTION TO 2.3.1, 2.4.1, 2.5.1

a) C_2H_3N

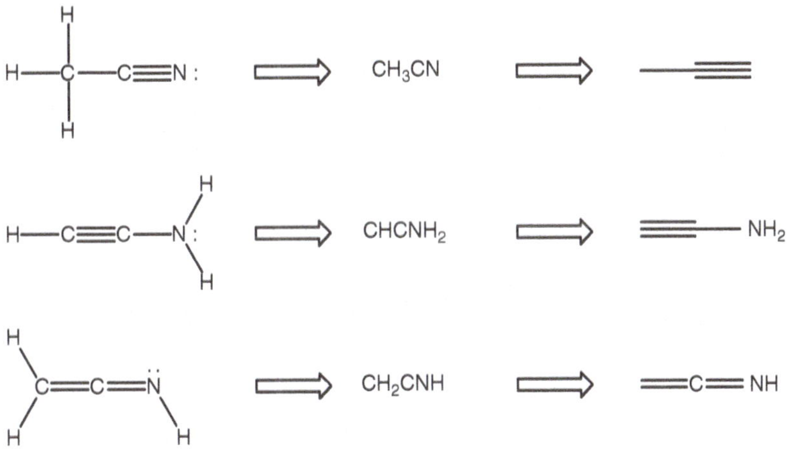

b) C_3H_6O

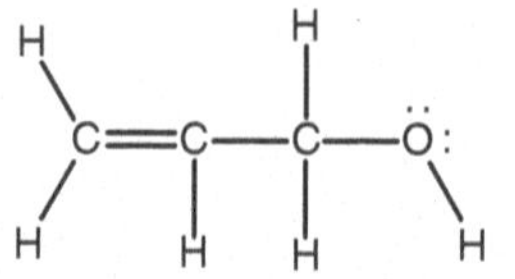

⟹ CH_2CHCH_2OH ⟹

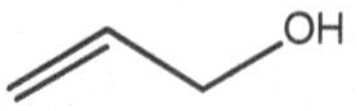

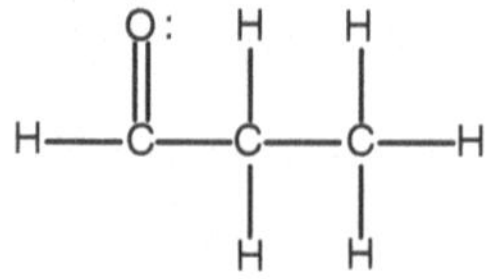

⟹ CH_3CH_2CHO ⟹

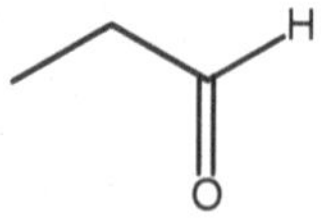

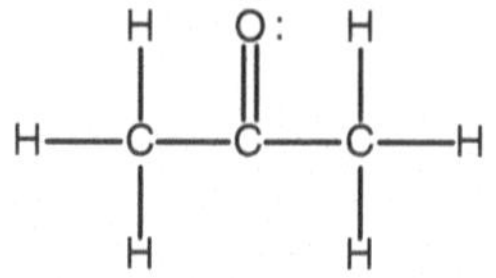

⟹ CH_3COCH_3 ⟹

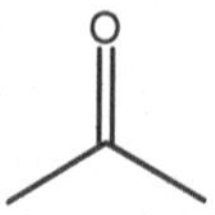

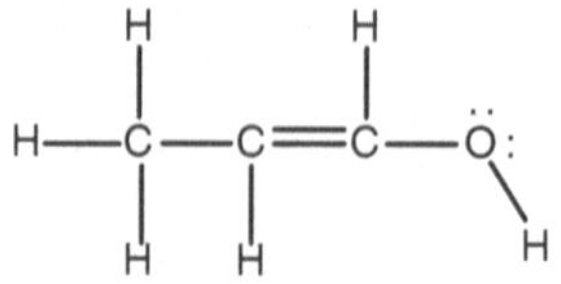

⟹ $CH_3CHCHOH$ ⟹

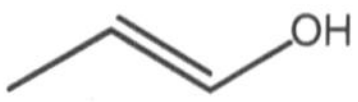

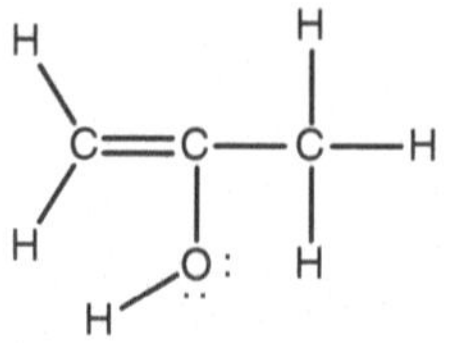

⟹ $CH_2C(OH)CH_3$ ⟹

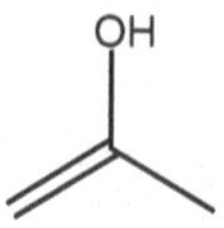

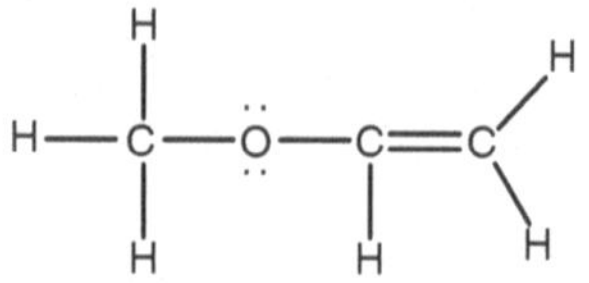

⟹ CH_3OCHCH_2 ⟹

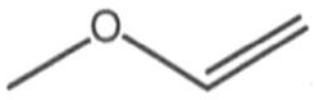

c) C_4H_9Cl

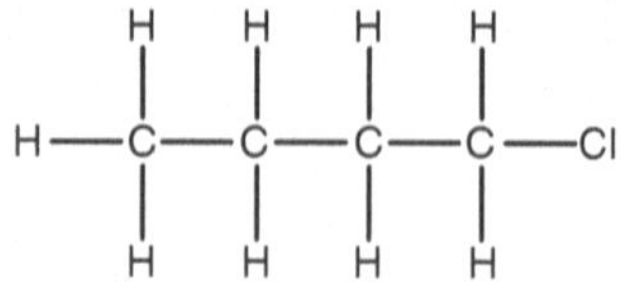

⟹ $CH_3CH_2CH_2CH_2Cl$ ⟹

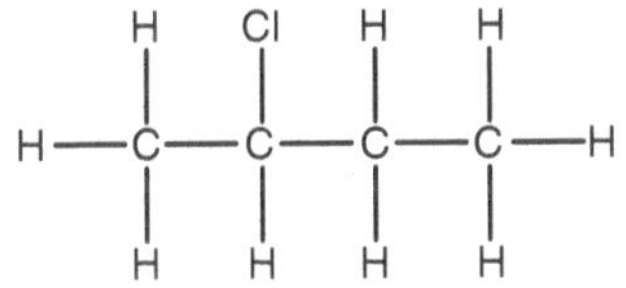

⇒ $CH_3CH(Cl)CH_2CH_3$ ⇒

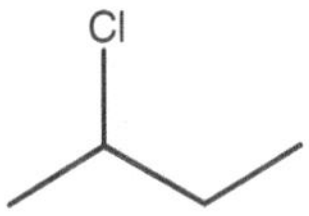

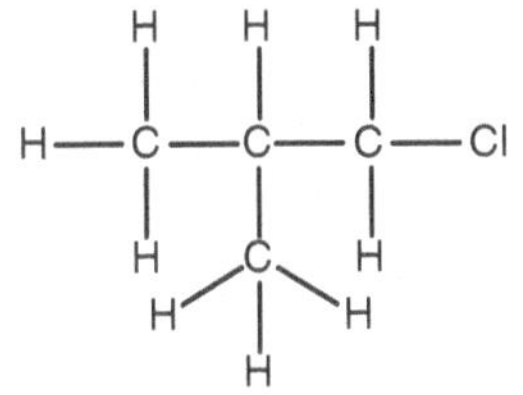

⇒ $(CH_3)_2CHCH_2Cl$ ⇒

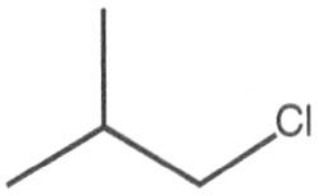

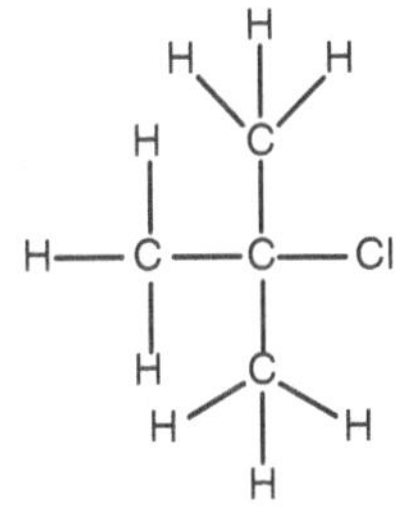

⇒ $(CH_3)_3CCl$ ⇒

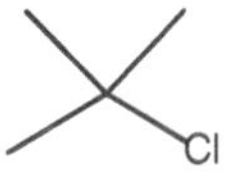

d) $\mathbf{C_5H_{12}}$

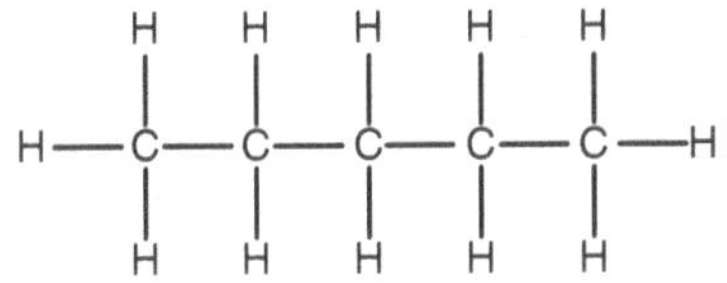

⇒ $CH_3CH_2CH_2CH_2CH_3$ ⇒

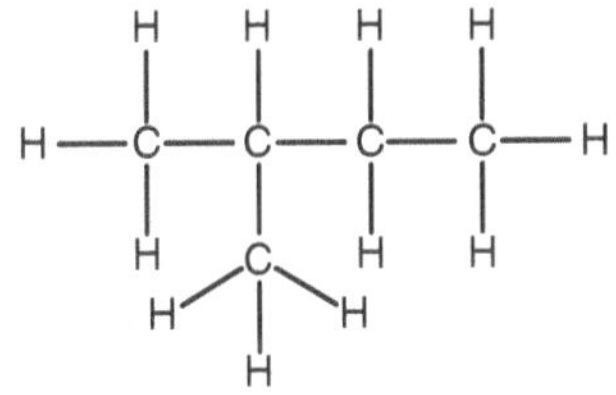

⇒ $(CH_3)_2CHCH_2CH_3$ ⇒

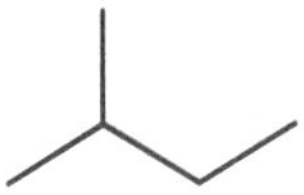

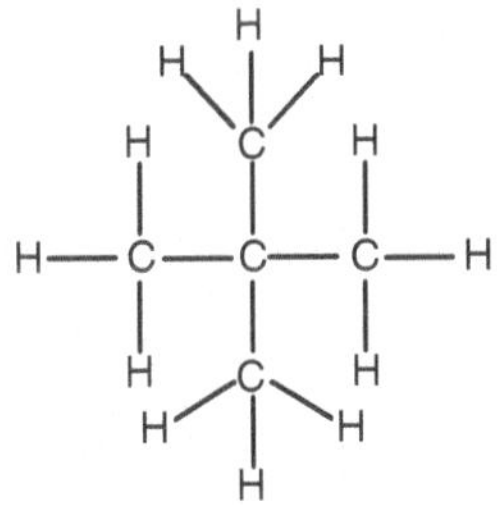

⇒ $(CH_3)_4C$ ⇒

SOLUTION TO 2.6.1

a) C2-C3 bond of $CH_3CH_2CH_2CH_2CH_3$.

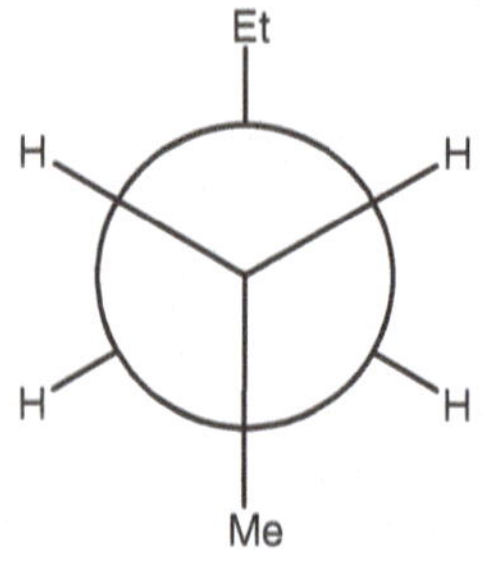

b) C5-C4 bond of 4-methyl heptane.

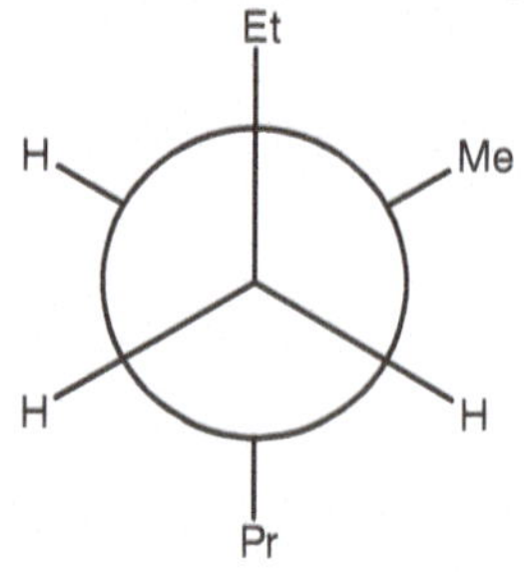

SOLUTION TO 2.6.2

a) C3-C4 bond of 3,4-dimethyl hexane

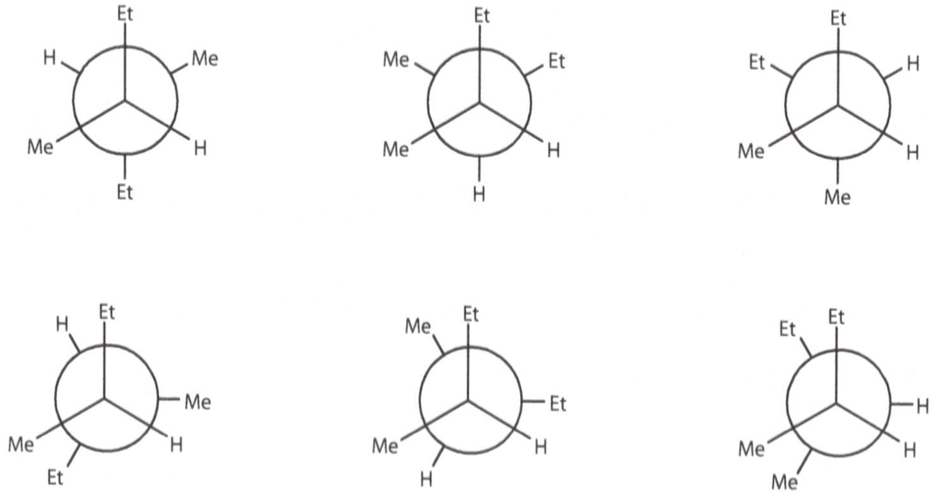

b) C2-C3 bond of 2-chloro-2-methylpentane

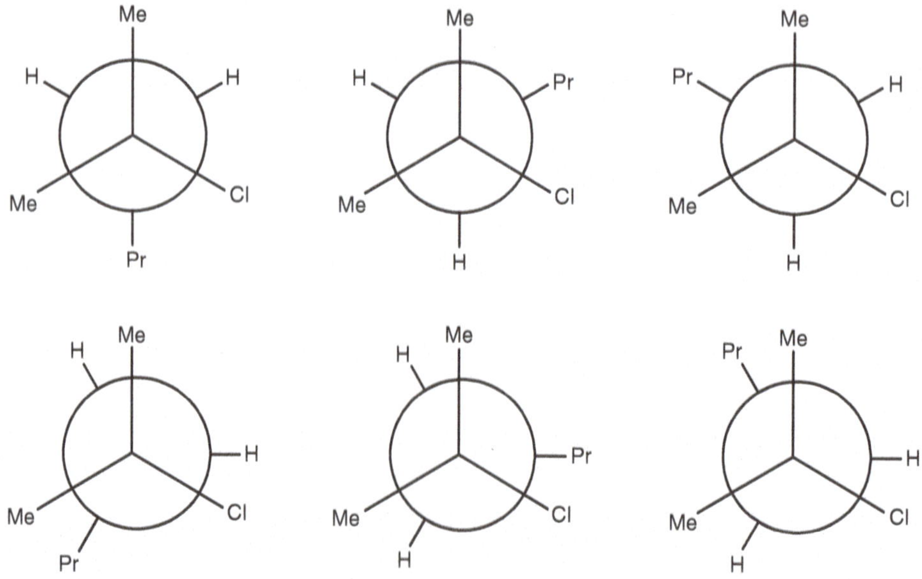

SOLUTION TO 2.7.1

a)

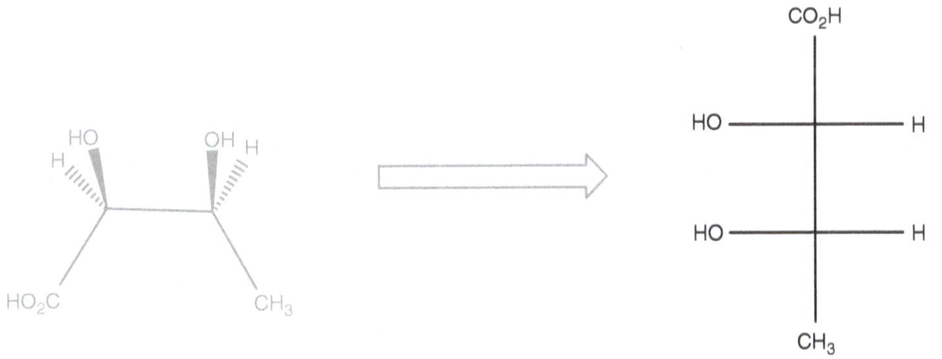

b) Draw the possible Fisher projections for all the 4 carbon aldoses (Erythrose and Threose)

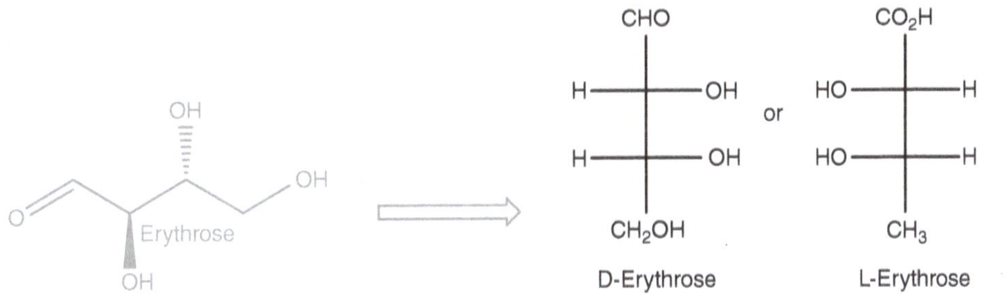

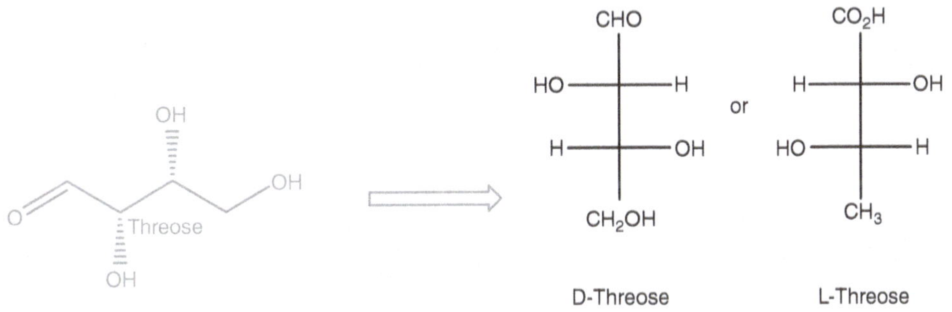

SOLUTION TO 2.8.1

1. Draw one of the chair conformations for the following molecules.

   a) 1-chloro-2-methylcyclohexane

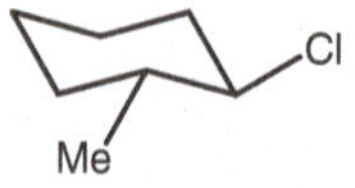

   b) Isopropylcyclohexane

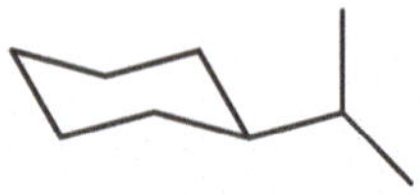

   c) 2-ethyl-5-methyl-1-propylcyclohexane

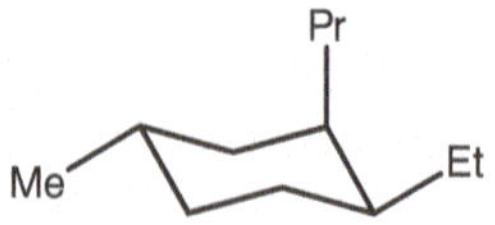

2. Draw ring flips for the following molecules

   a) t-Butylcyclohexane

   b) 1-cholo-2-fluoro-cyclohexane

   c) 1-Bromo-3-methyl-cyclohexane

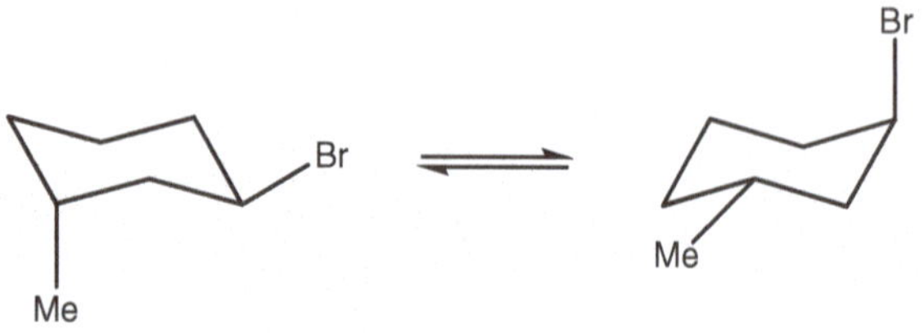

SOLUTION TO 3.3.1

a)

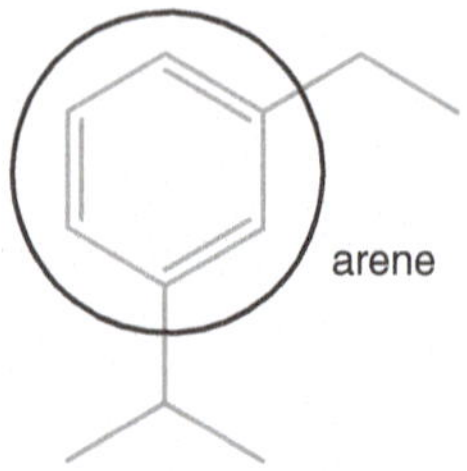

b)

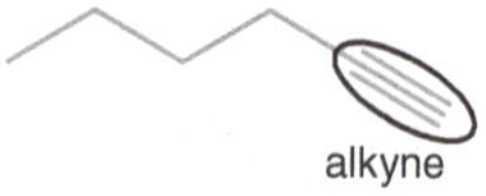

c)

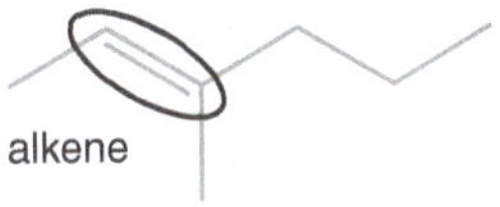

d)

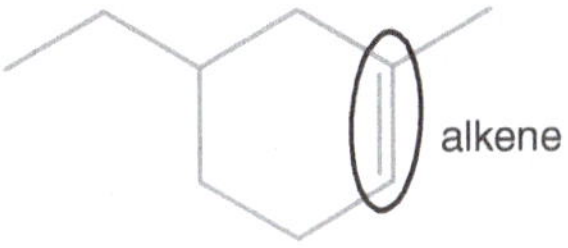

e)

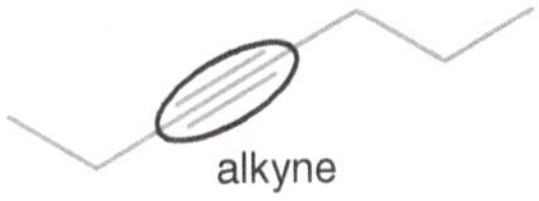

f)

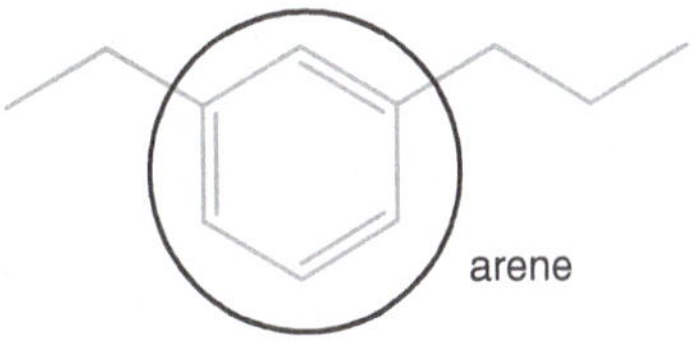

SOLUTION TO 3.4.1

a)

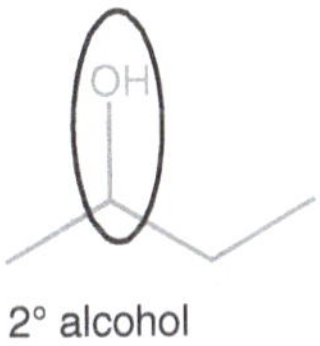

b)

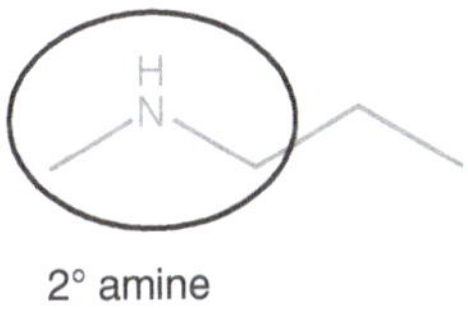

c)

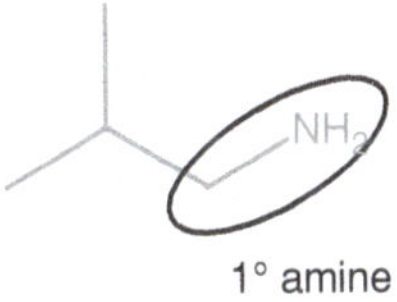

d)

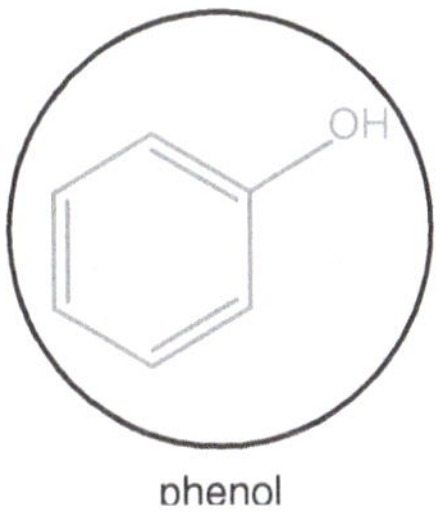

e)

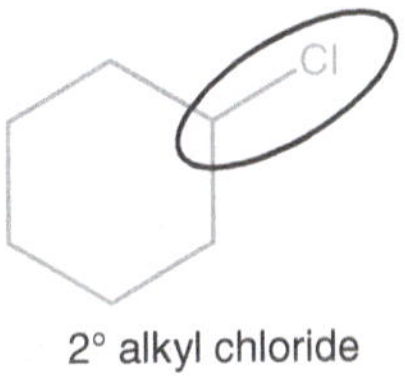

f)

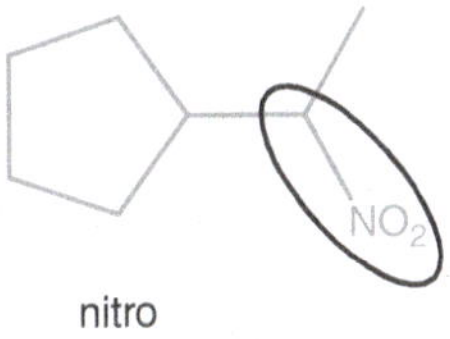

g)

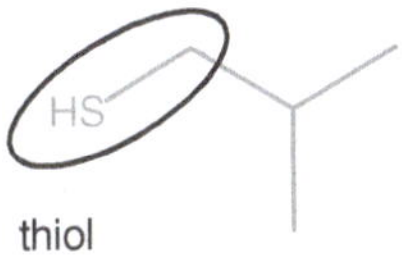

h)

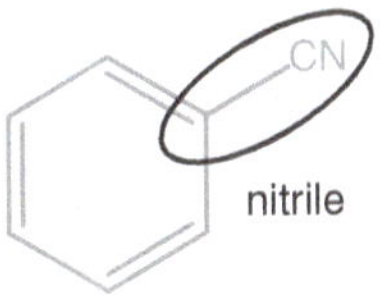

i)

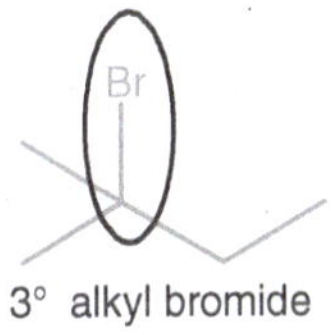

j)

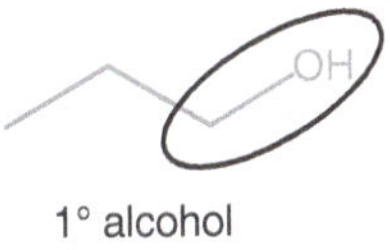

k)

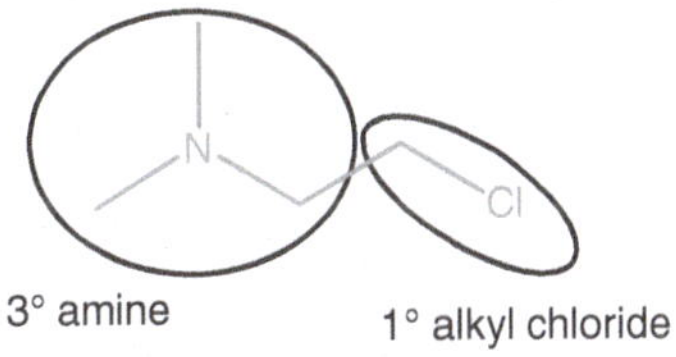

l)

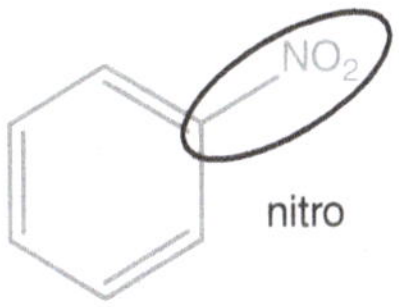

SOLUTION TO 3.5.1

a)

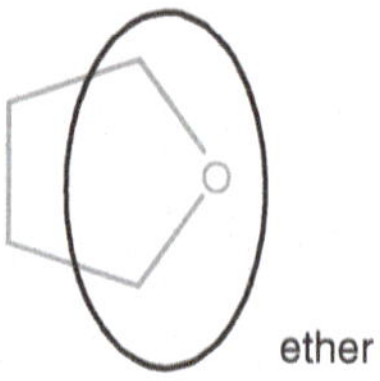

b)

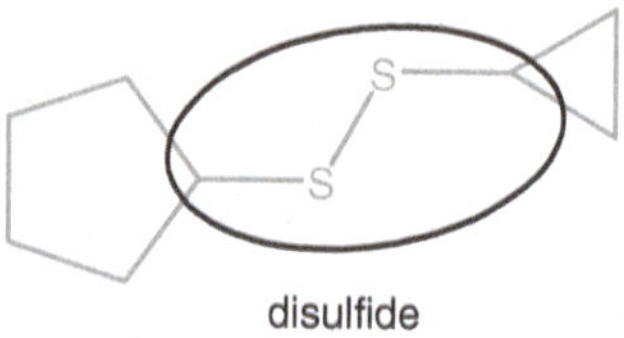

c)

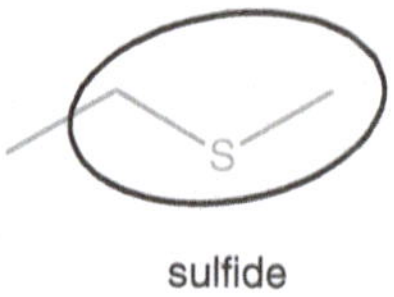

d)

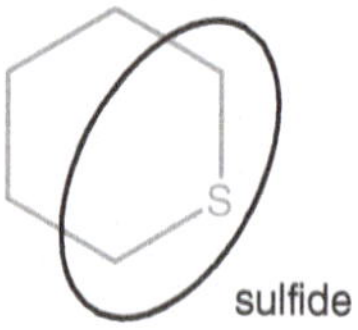

e)

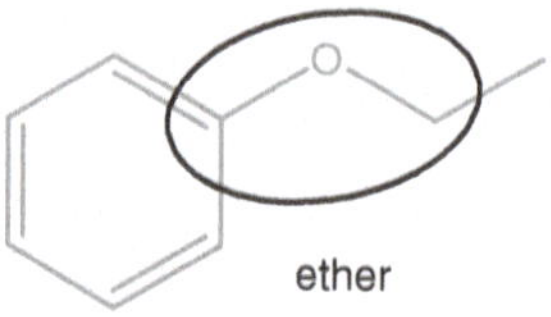

f)

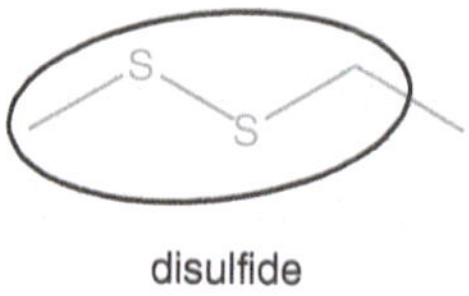

SOLUTION TO 3.6.1

a)

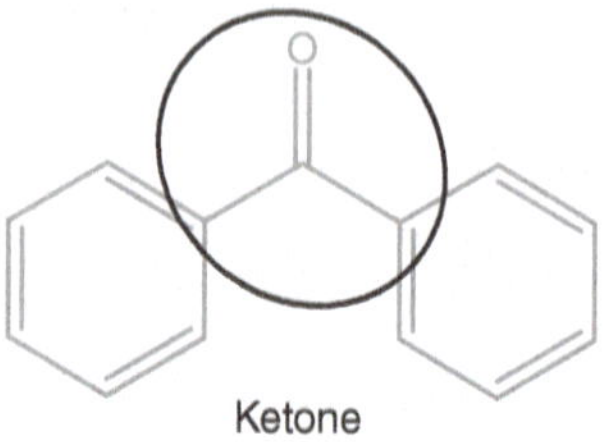

b)

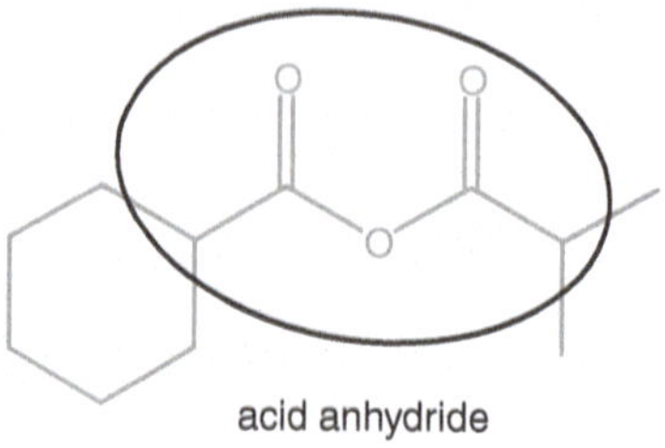

c)

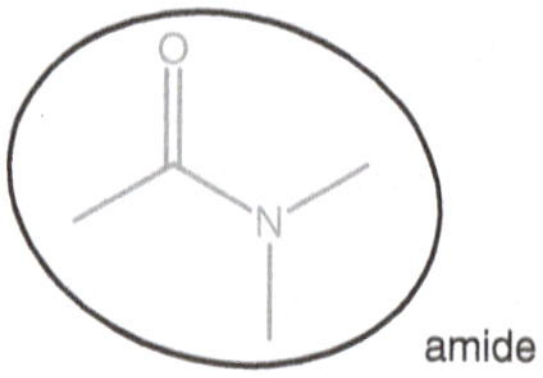

d)

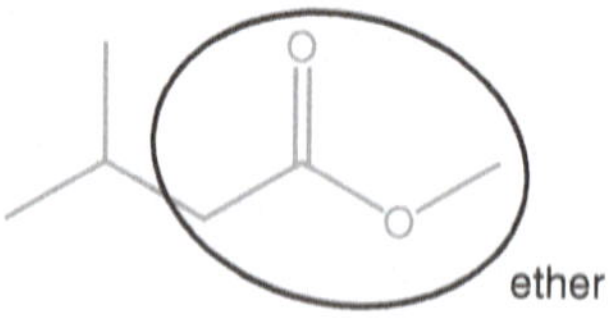

e)

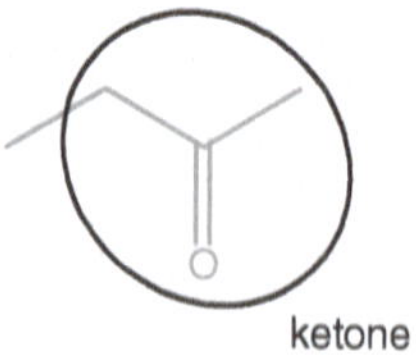

f)

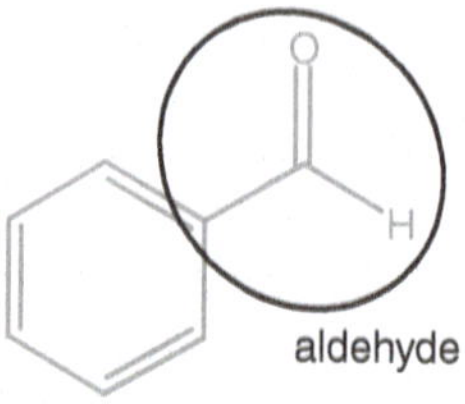

g)

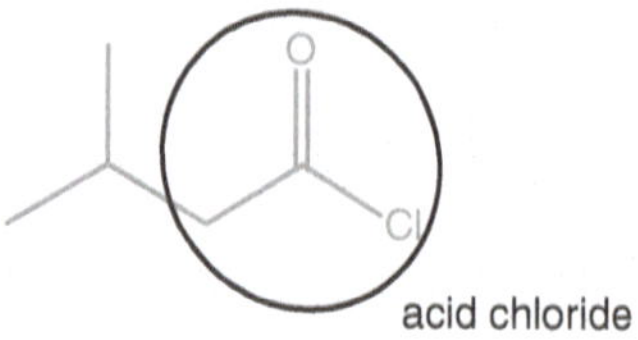

h)

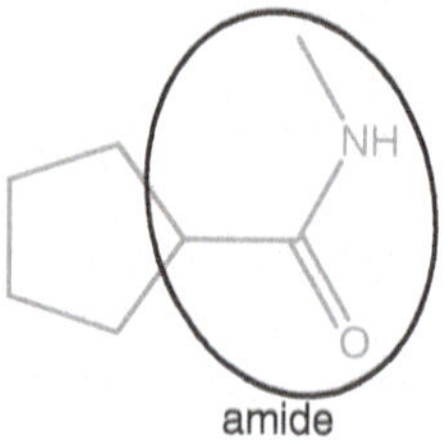

i)

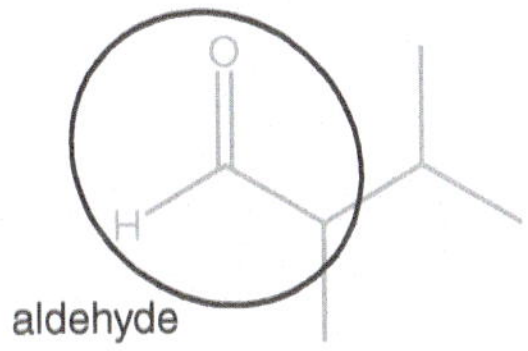

j)

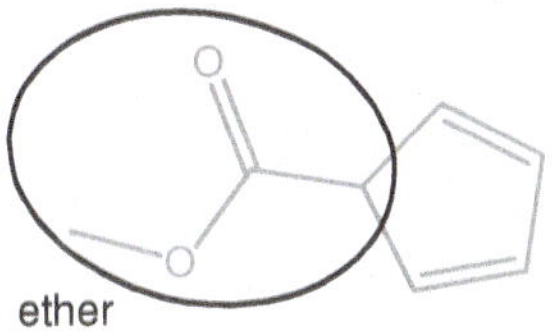

SOLUTION TO 4.2.1

I highlighted the parent chain and indicated carbon 1 for numbering.

a)

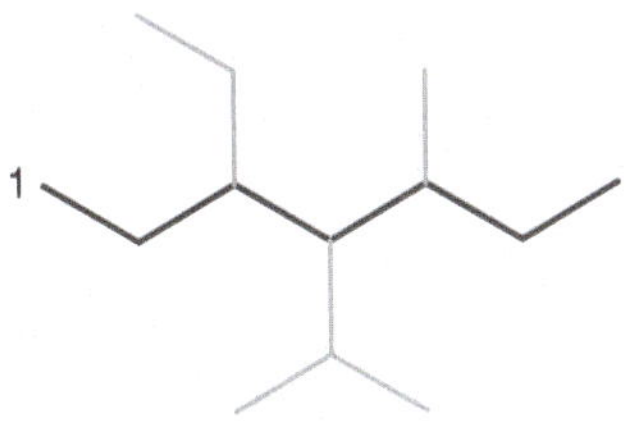

3-ethyl-4-isopropyl-5-methylheptane
Note: if the numbers are the same, then the one that comes first in the alphabet gets the lowest number

b)

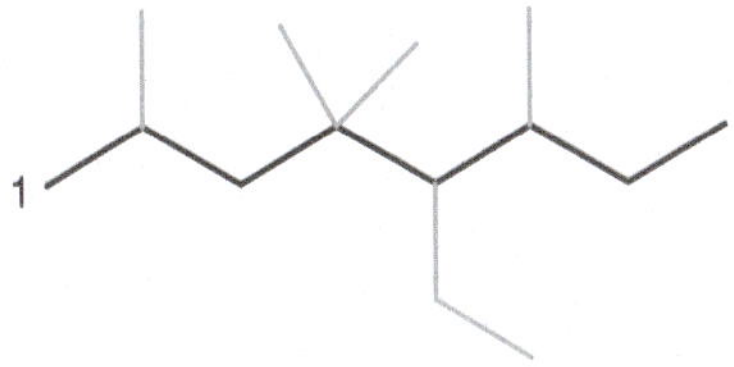

5-ethyl-2,4,4,6-tetramethyloctane

c)

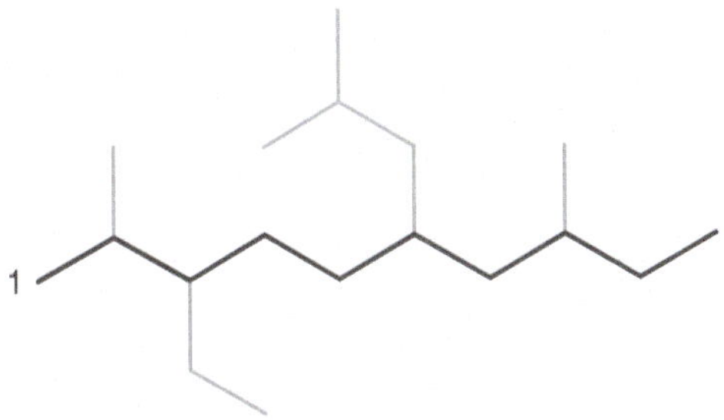

3-ethyl-6-isobutyl-2,8-dimethyldecane

d)

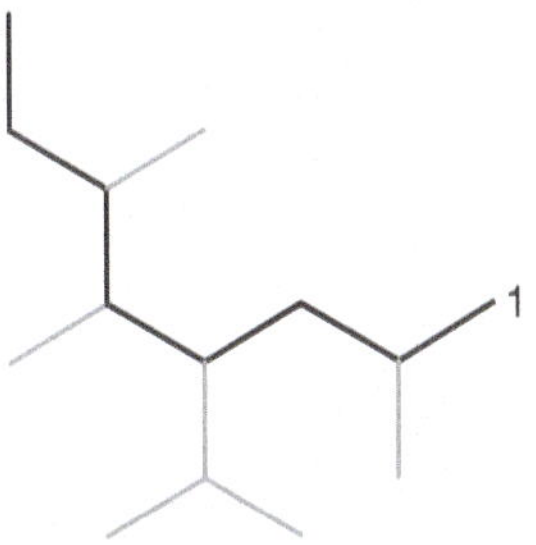

4-isopropyl-2,5,6-trimethyloctane

e)

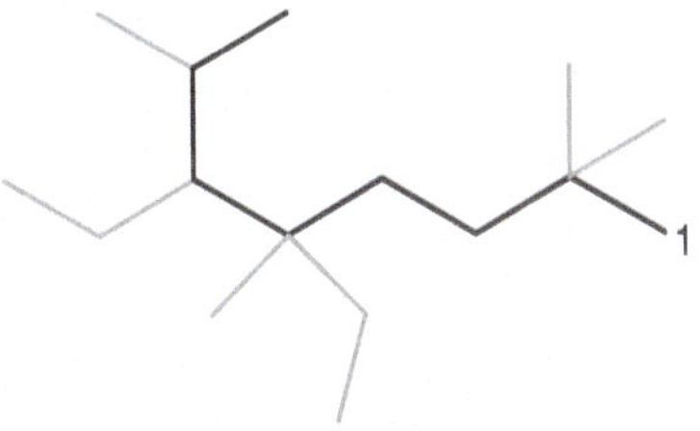

5,6-diethyl-2,2,5,7-tetramethyloctane

f)

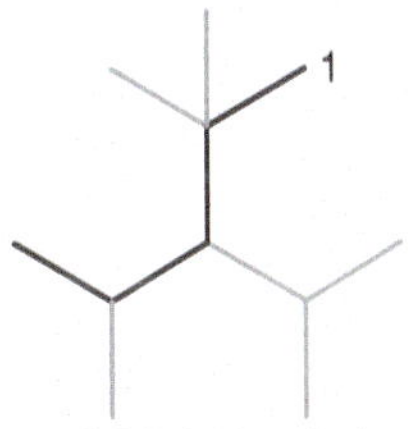

3-isopropyl-2,2,4-trimethylpentane

SOLUTION TO 4.2.2

a) 4-Ethyl-2,4,7-trimethyloctane

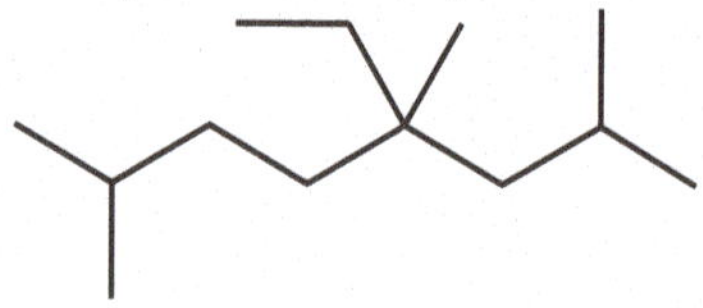

b) 3,4-Diethyl-2,2,3,5-tetramethylhexane

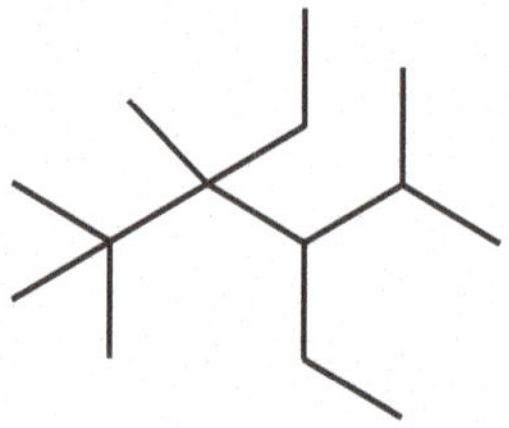

c) 2,2-Dimethylheptane

d) 8-tert-Butyl-3,11-diethyl-5-propyltridecane

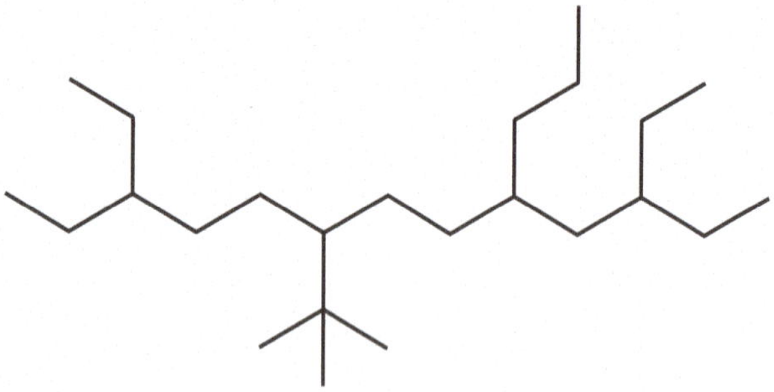

e) 4-Elhyl-5-isobutyl-6-isopropyl-2,2-dimethyldecane

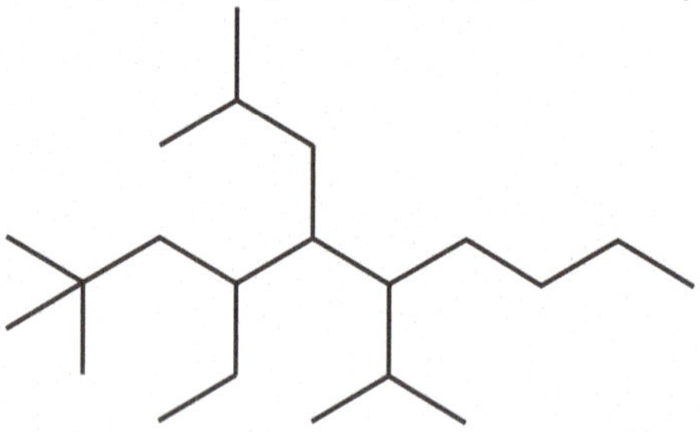

SOLUTION TO 4.3.1

a)

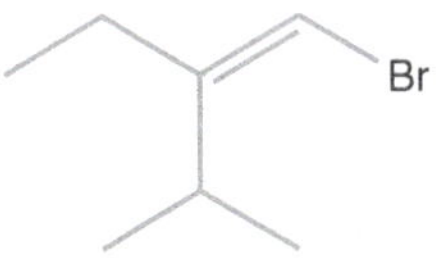

(*Z*)-3-(bromomethylene)-2-methylpentane

b)

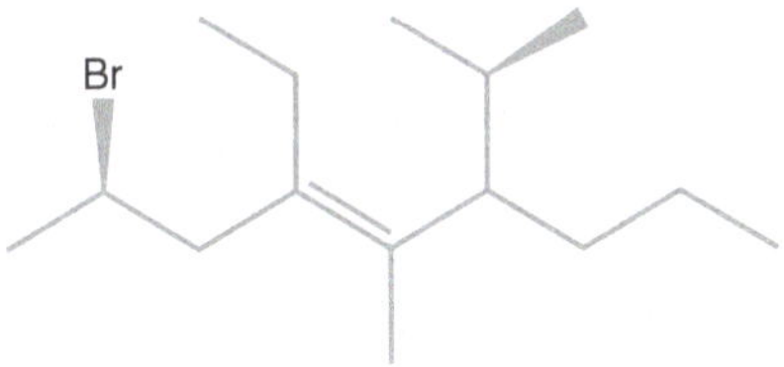

(2*R*,*E*)-2-bromo-4-ethyl-6-isopropyl-5-methylnon-4-ene

c)

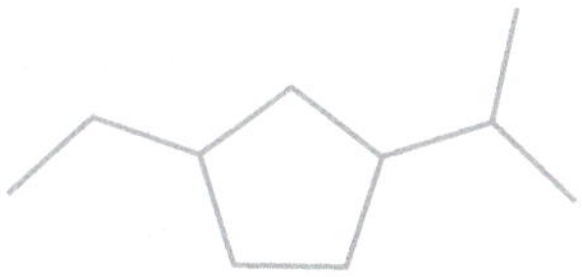

1-ethyl-3-isopropylcyclopentane

d)

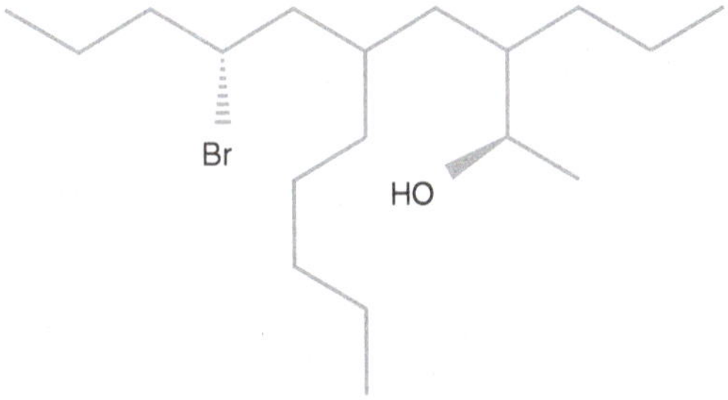

(2*R*,7*R*)-7-bromo-5-pentyl-3-propyldecan-2-ol

e)

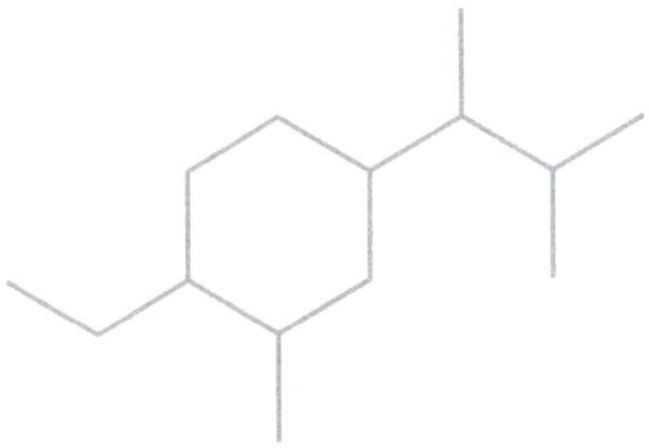

1-ethyl-2-methyl-4-(3-methylbutan-2-yl)cyclohexane

f)

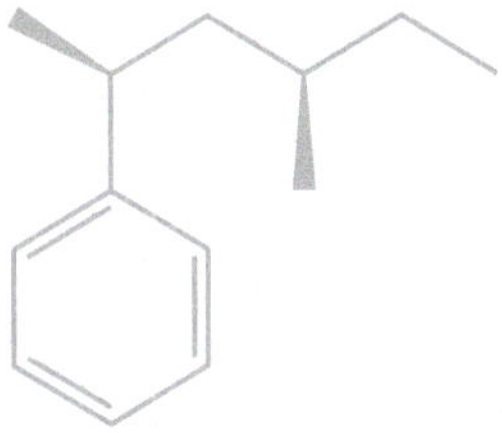

((2*S*,4*R*)-4-methylhexan-2-yl)benzene

g)

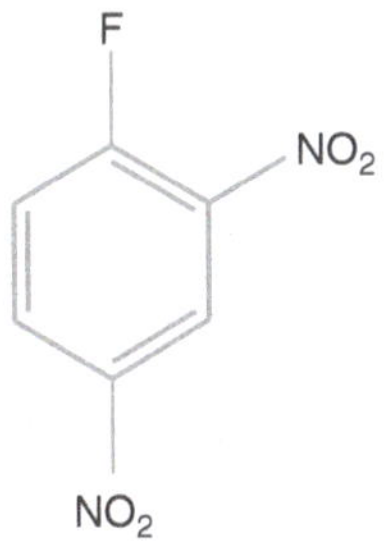

1-fluoro-2,4-dinitrobenzene

h)

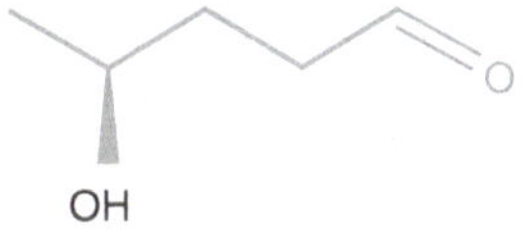

(*S*)-4-hydroxypentanal

SOLUTION TO 4.3.2

a) R-2-bromo-3-hexanone

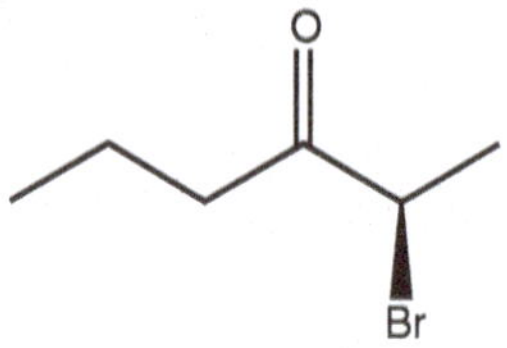

b) N,N,4-trimethylpentanamide

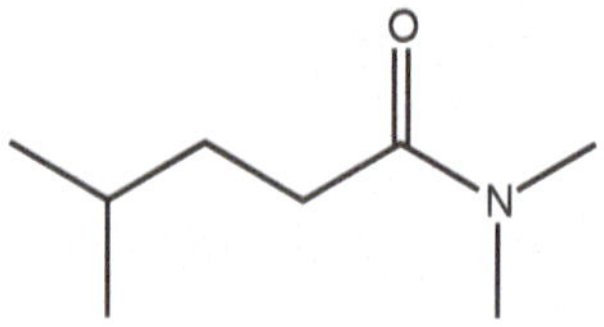

c) 1-Methyl-4-(methylethyl)cyclohexane

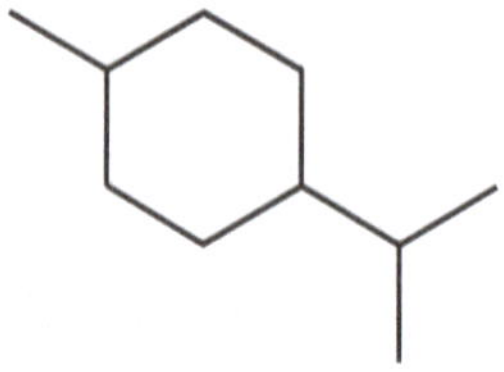

d) 3-aminohexane

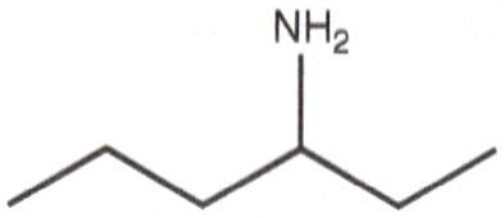

e) 3-chlorobenzaldehyde

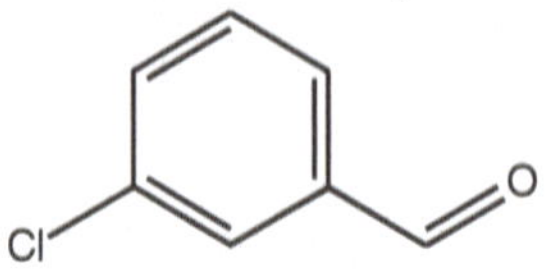

SOLUTION TO 5.3.1

a)

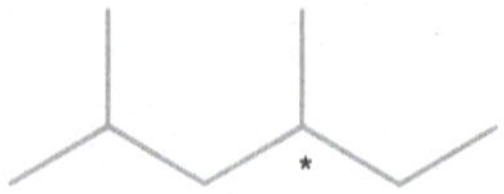

b)

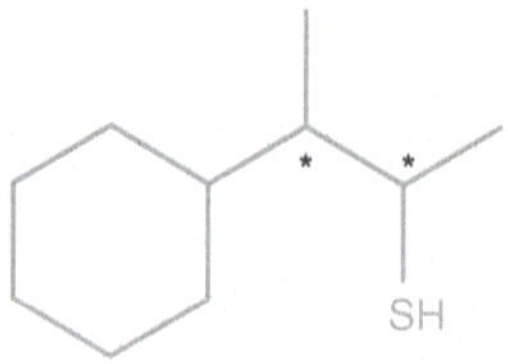

c)

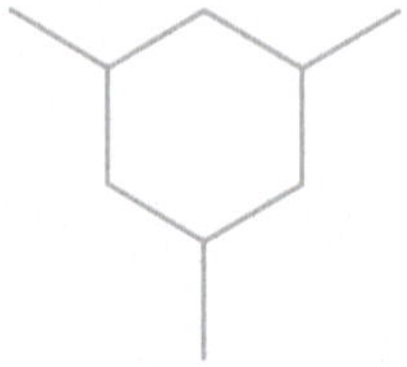

d)

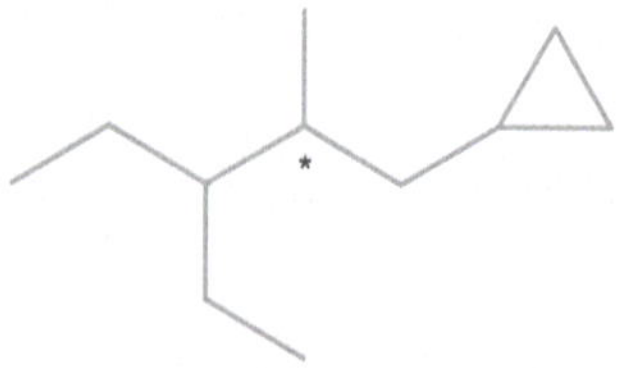

e)

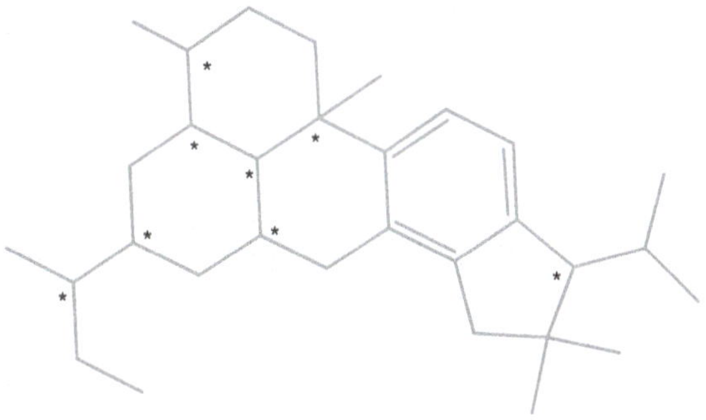

f)

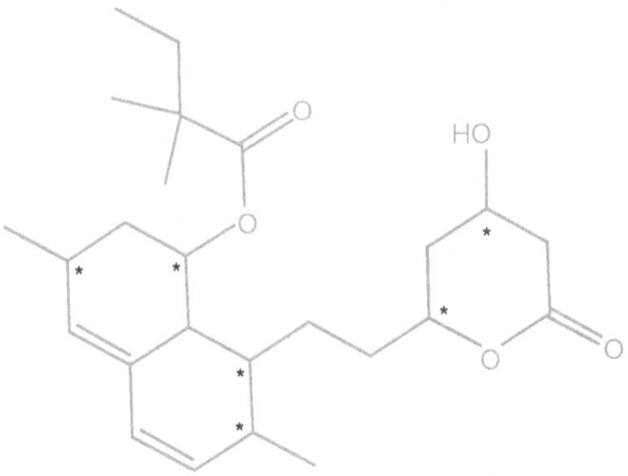

g)

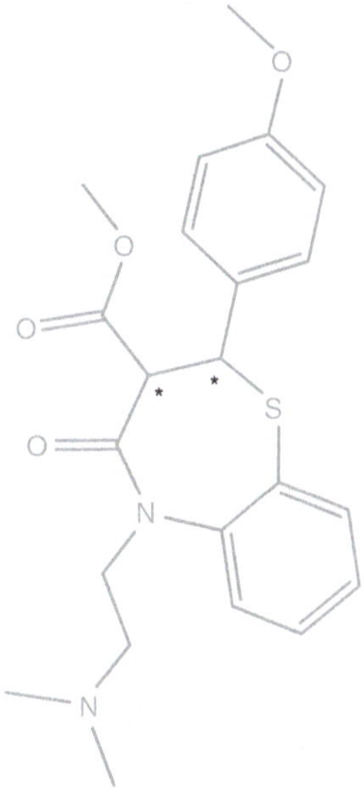

SOLUTION TO 5.3.2

a)

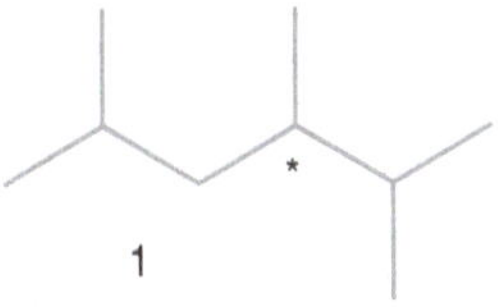

1

b)

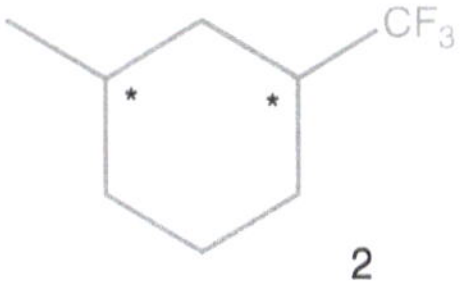

2

c)

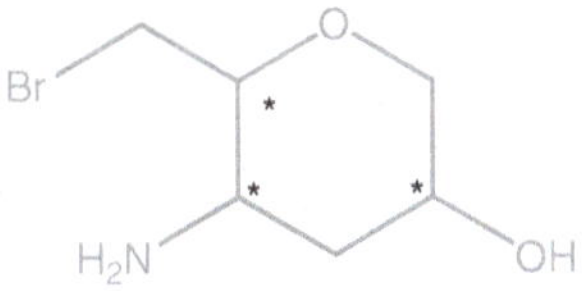

3

d)

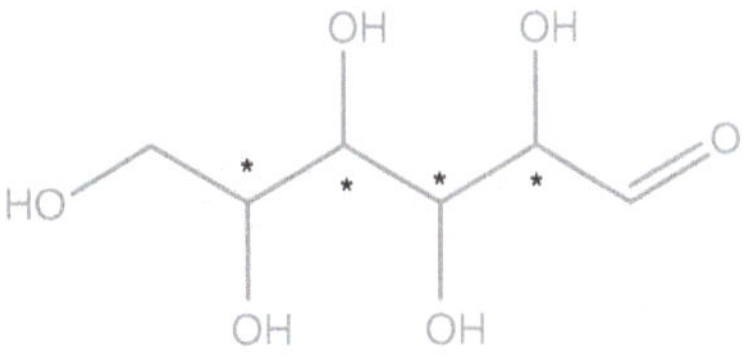

4

e)

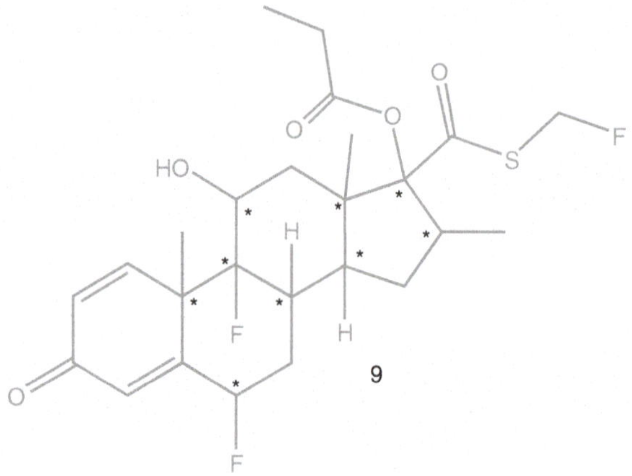

9

f)

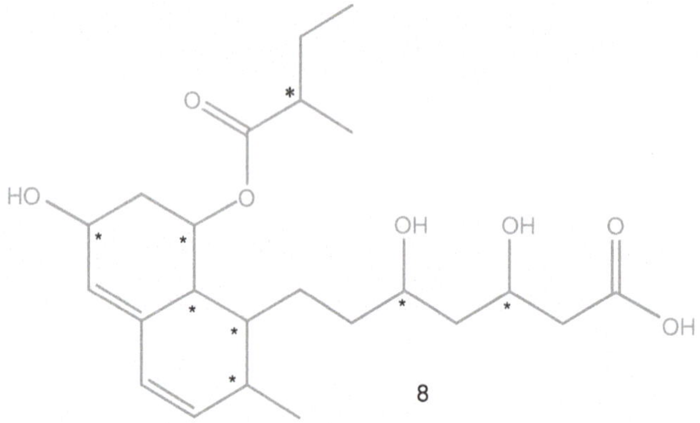

8

SOLUTION TO 5.3.3

a)

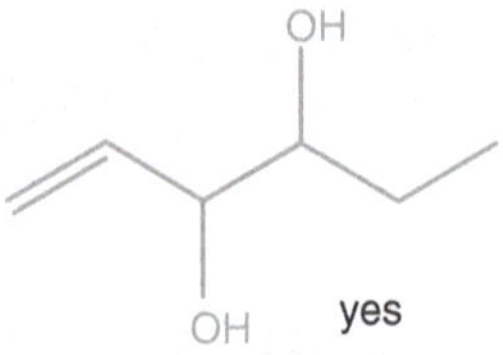

yes

b)

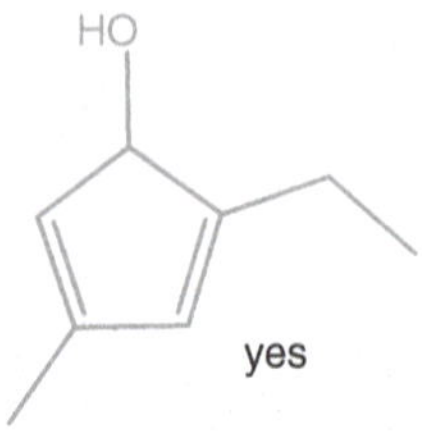

yes

c)

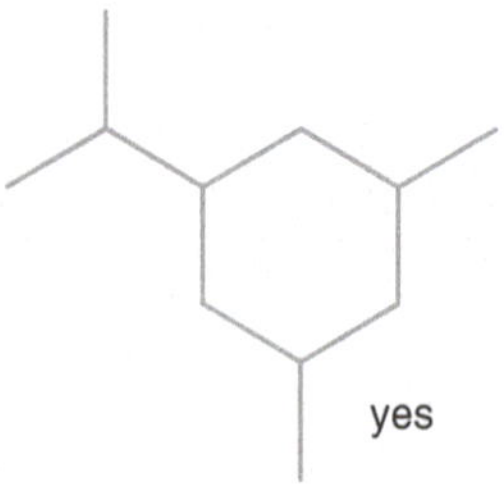

yes

d)

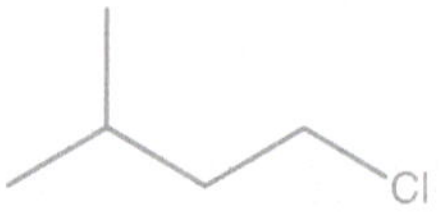

no

e) 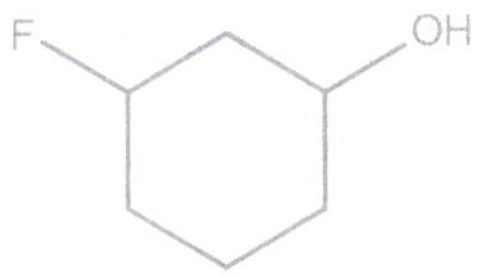

yes

f) 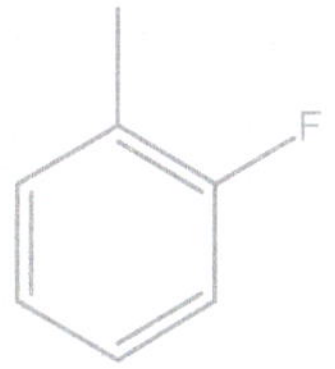

no

g) 

no

h) 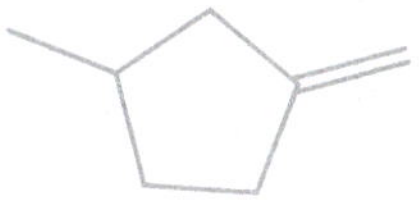

yes

SOLUTION TO 5.4.1

a) 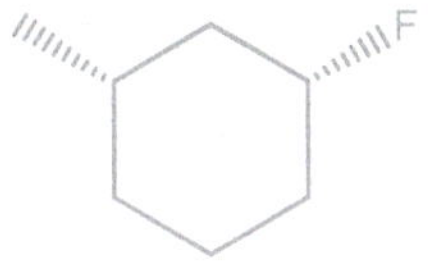

cis

b) 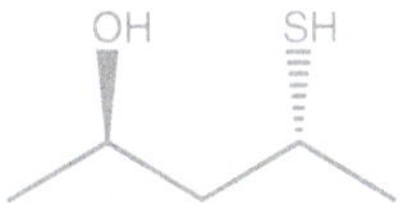

trans

c) 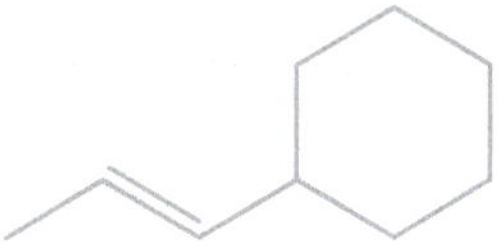

trans

d) 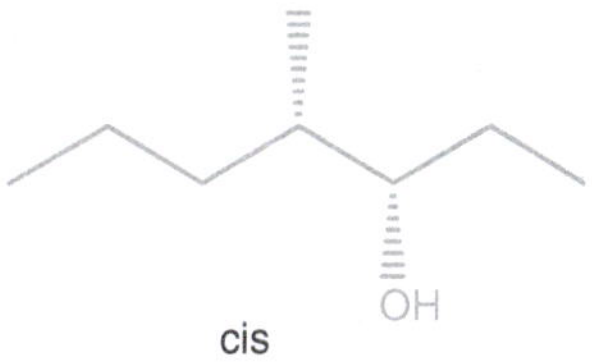

cis

e) 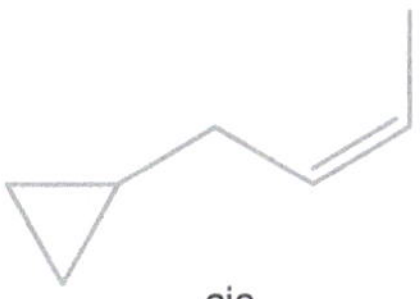

cis

f) 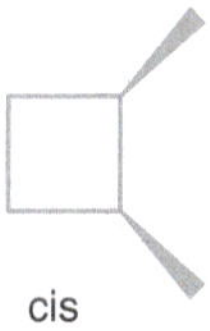

cis

g) 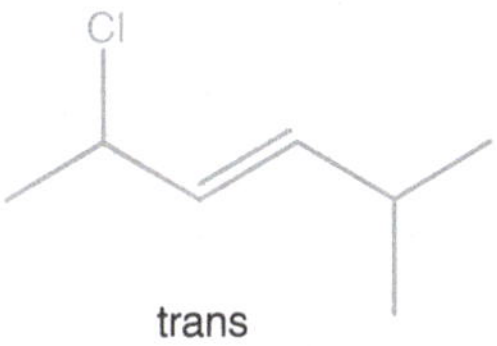

trans

h) 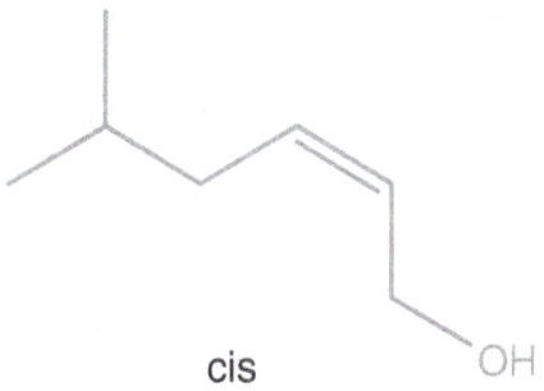

cis

SOLUTION TO 5.5.1

a)

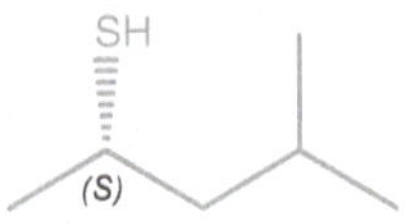

b)

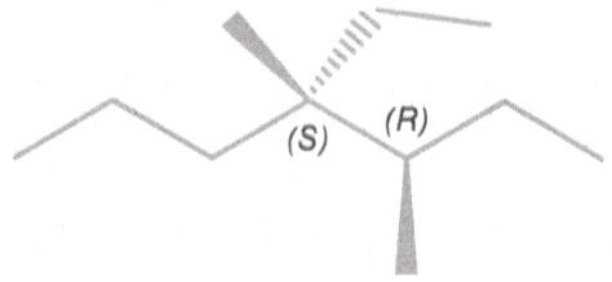

c)

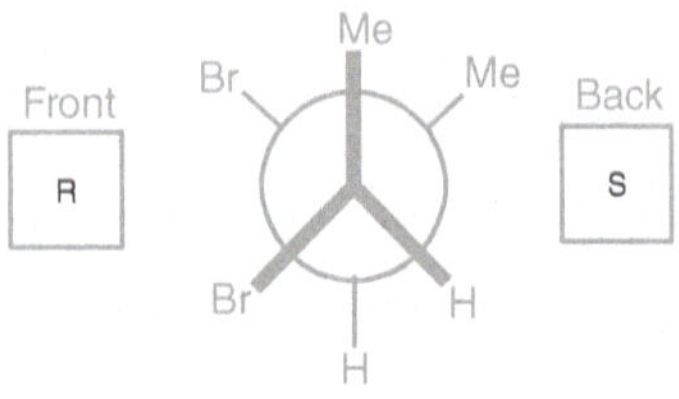

d)

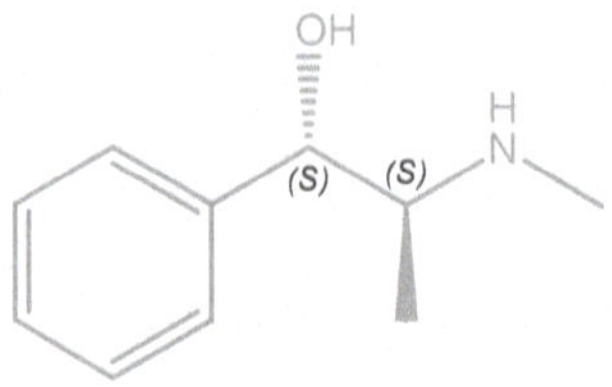

e)

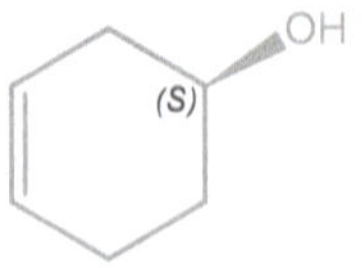

f)

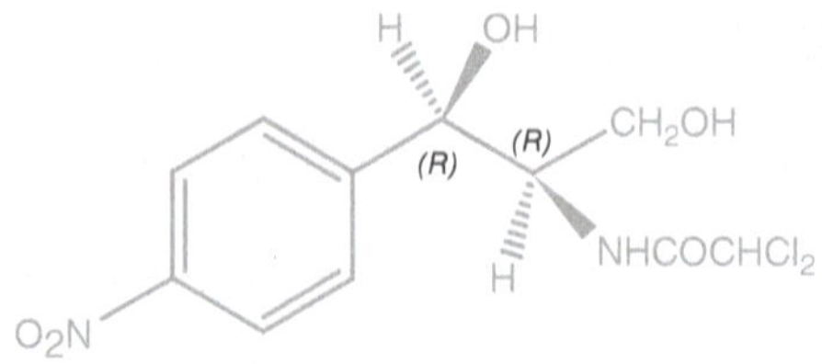

g)

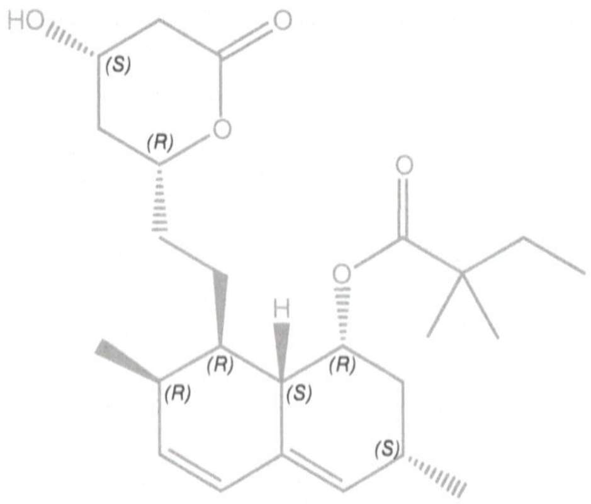

h)

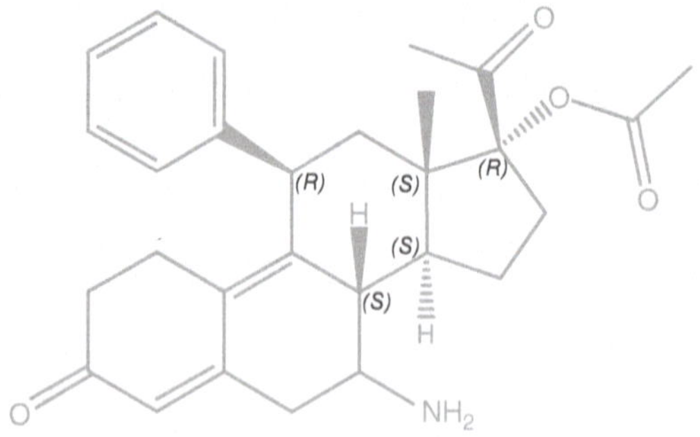

SOLUTION TO 5.5.2

a) (*R*)-2,3-Dimethyloctane

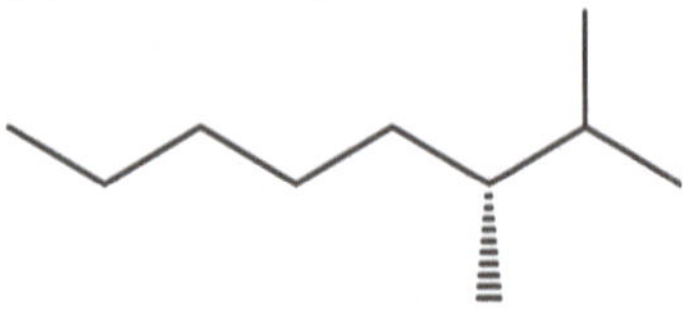

b) (4R,5S)-4-ethyl-2,4,5-trimethylheptane

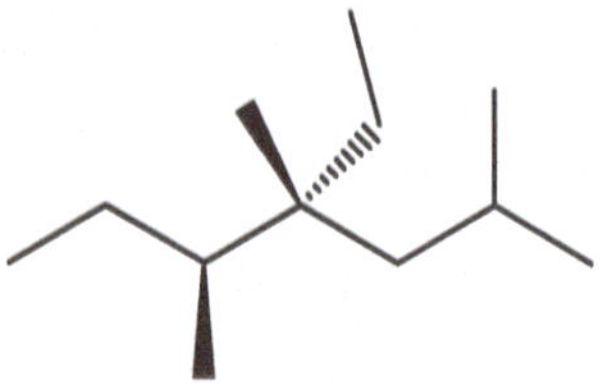

c) (S)-3-chloro-2-methylpentane

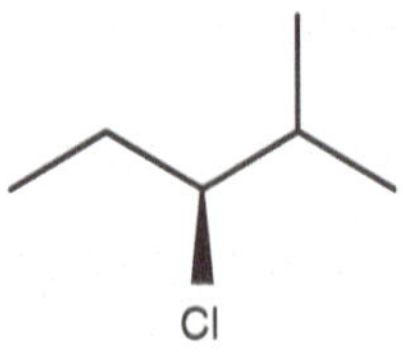

d) (1R,3*R*)-1-isopropyl-3-methylcyclohexane

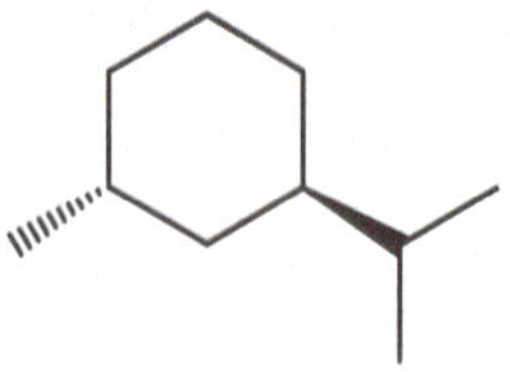

e) (*S*)-3-ethyl-2,5-dimethylhexane

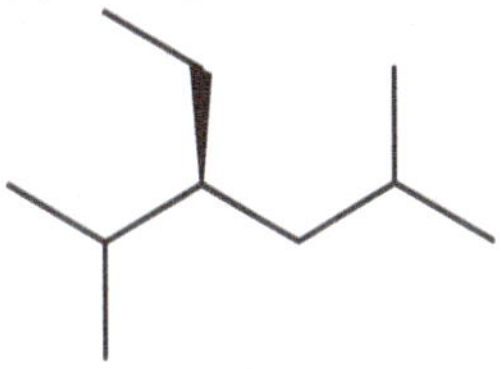

f) (2*R*,4*R*)-2-chloro-4-fluoro-4-methylhexane

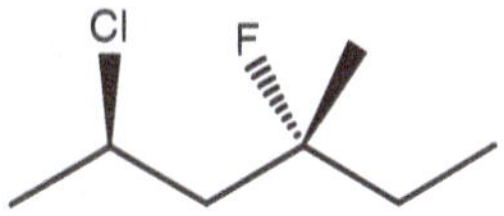

SOLUTION TO 5.6.1

a)

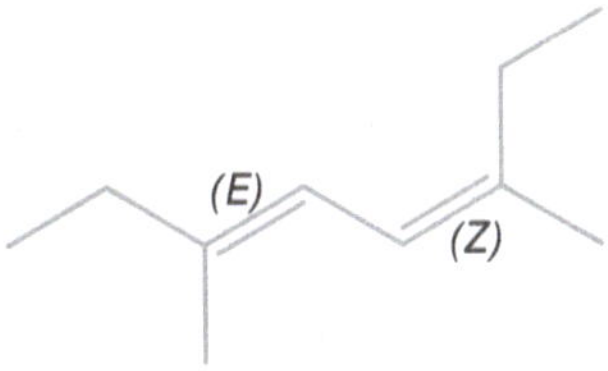

b)

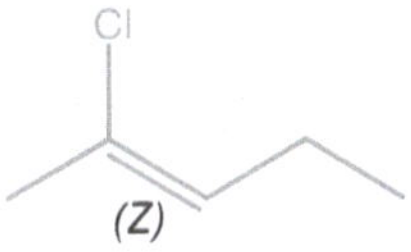

c)

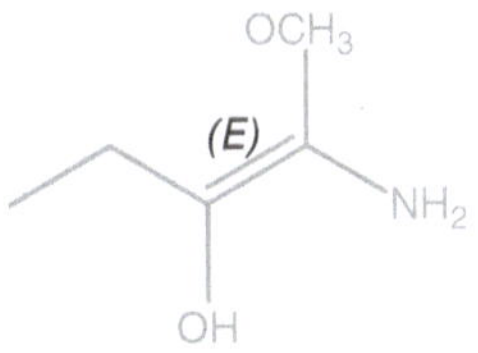

d)

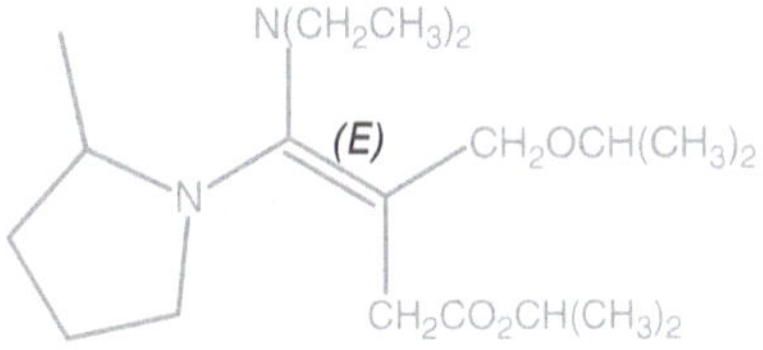

e)

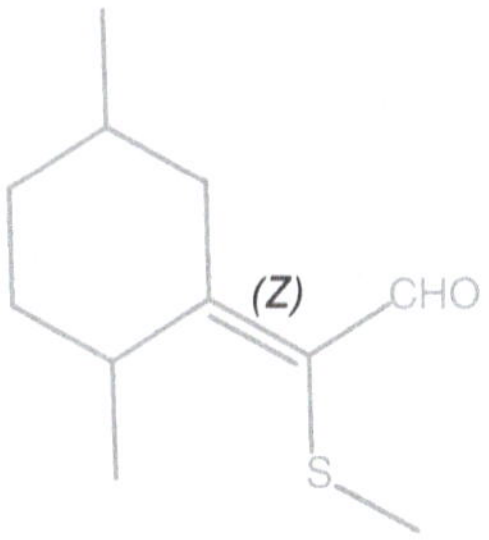

f)

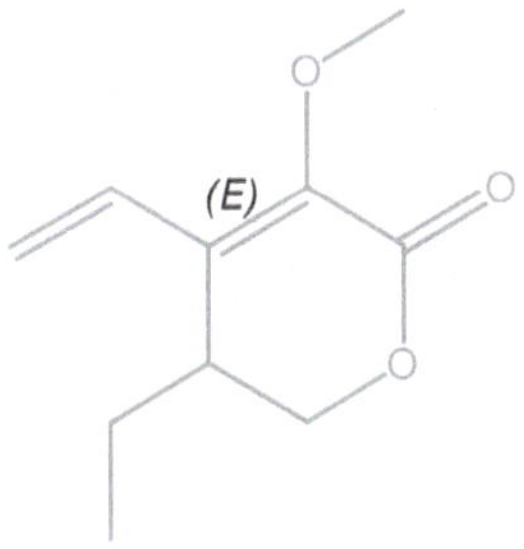

g)

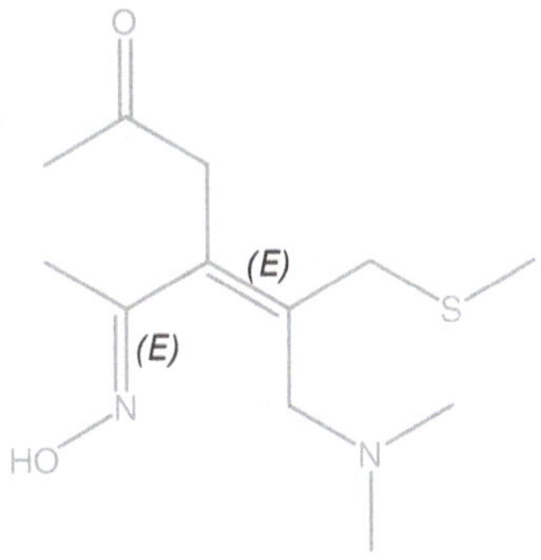

h)

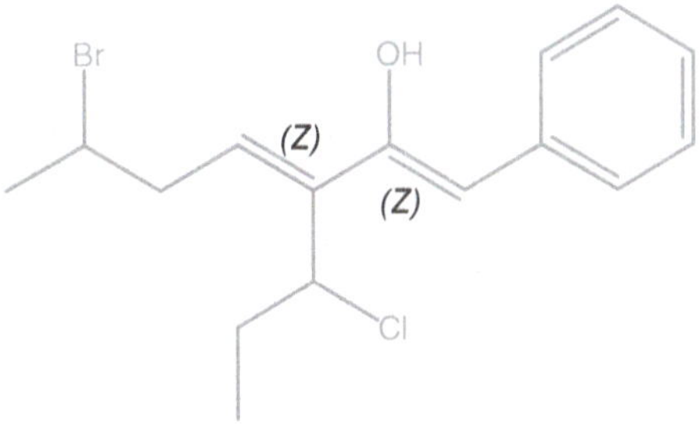

SOLUTION TO 5.7.1

a)

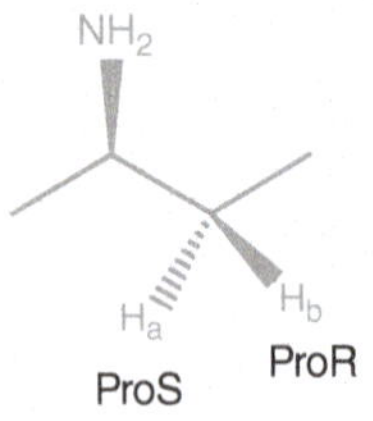

b)

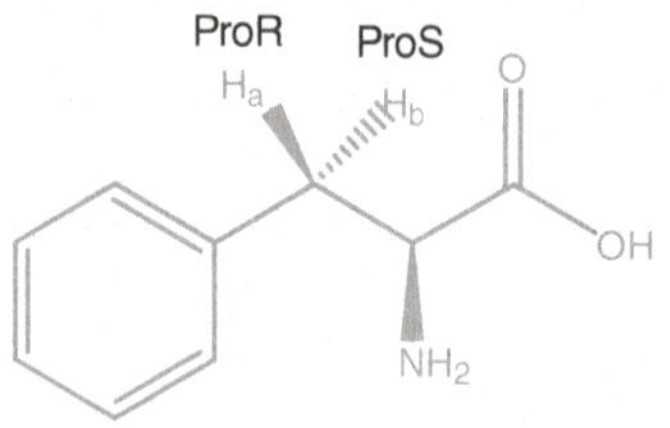

c)

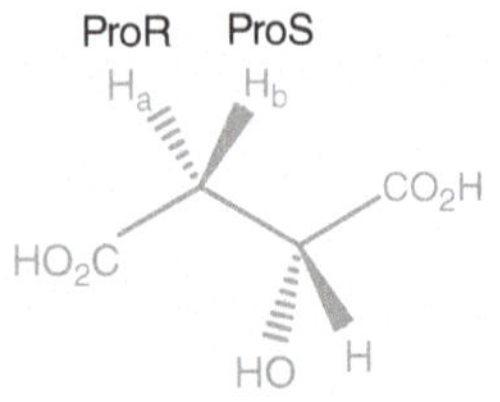

d)

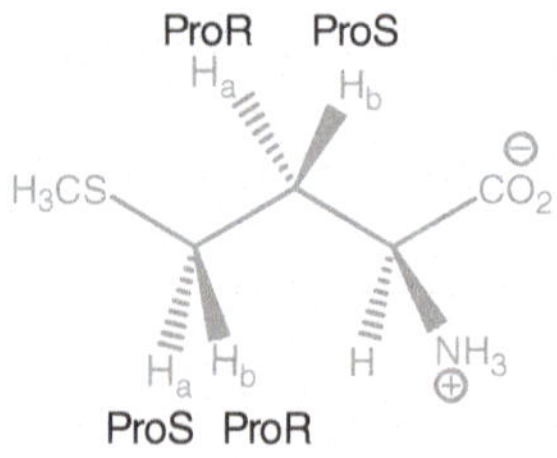

e)

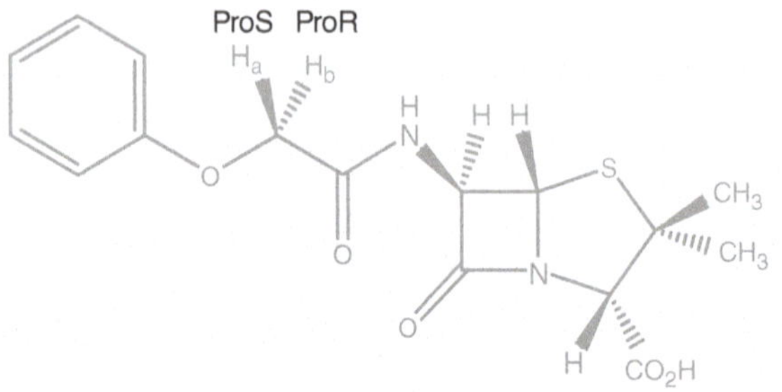

SOLUTION TO 5.7.2

a)

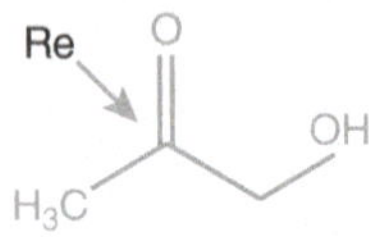

b)

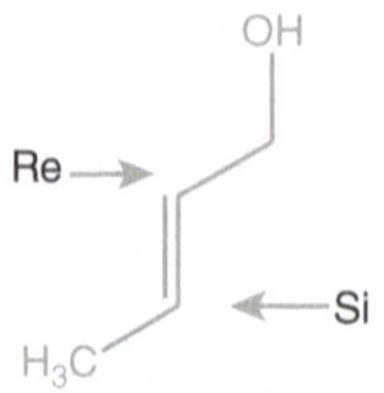

c)

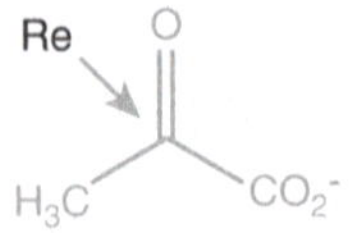

d)

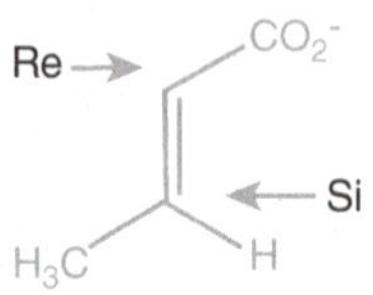

e)

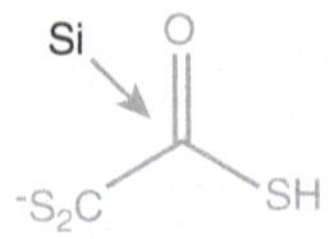

f)

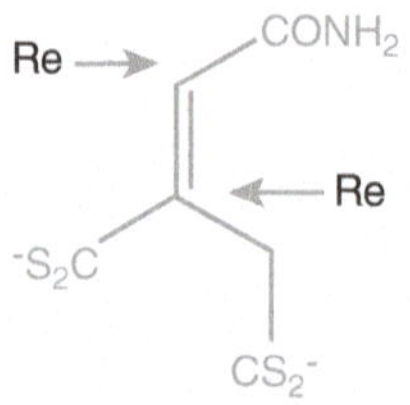

SOLUTION TO 5.8.1

a)

Constitutional isomer

b)

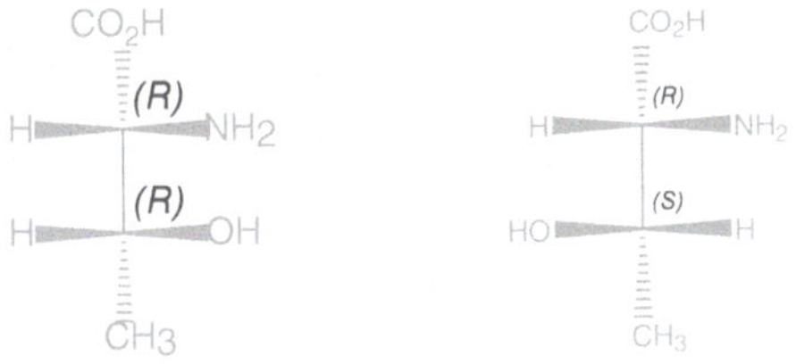

Diastereomer

c)

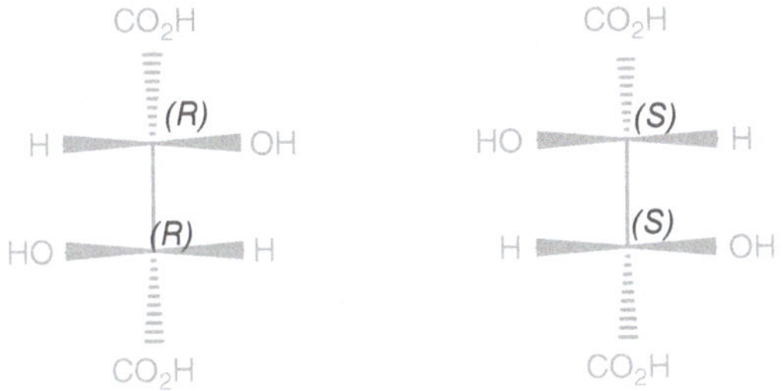

Enantiomers

d)

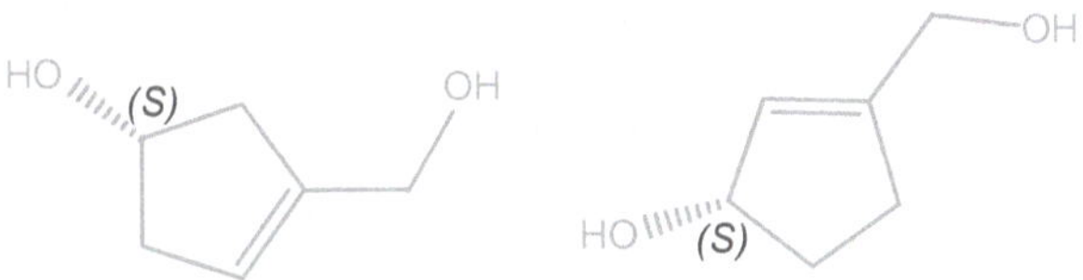

Constitutional isomer

e)

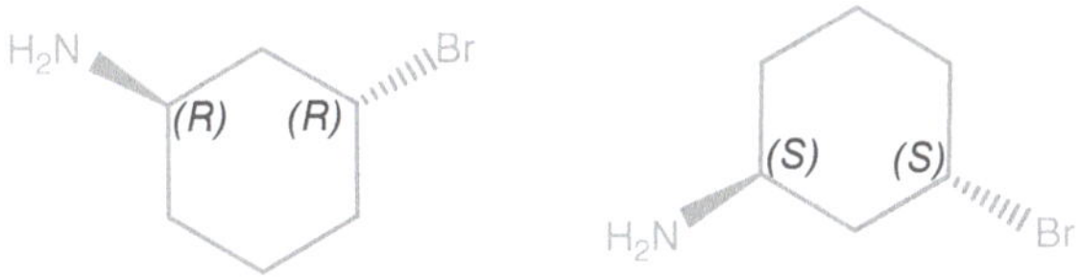

Enantiomer

f)

Enantiomer

g)

Meso